MARINE GEOMORPHOLOGY:

2nd EDITION

written by

N. Christian Smoot

ISBN: Softcover 978-1-4257-5541-6

This book was printed in the United States of America.

Book Designer: EJ Nemenzo

To order additional copies of this book, contact:
Xlibris Corporation
1-888-795-4274
www.Xlibris.com
Orders@Xlibris.com

Table of Contents

FOREWORD and ACKNOWLEDGEMENTS

Several years ago I realized there was a need for a marine geomorphology book. As a teacher of new hires in the Bathymetry Division of the Naval Oceanographic Office, I wrote and updated that training manual four times between 1980 and 1997. After I retired in 1998, I decided to do just that, using the materials that I had learned about and taught over the years.

My experience did not include seismic interpretation or terrestrial surveying. For that expertise, I have asked several of those more knowledgeable than I add the deeper thought to the book. Where they have been consulted, their names appear beside the chapter title.

Ismail is a specialist in the Himalayan region, magma floods, and world-wide tectonics, having spent over 25 years at the Wadia Institute of Himalayan Geology as a field hand, collecting samples and analyzing them. He got the call to return home, and Ismail is now the head of the geology and geophysics department at the University of Kashmir. Dong is a consulting geologist who is the manager of RAAX International in Australia. His early years were spent collecting ocean floor samples and analyzing them. When Dong was working for the Australian Bureau of Mineral Resources (now Geoscience Australia), he prepared a manuscript presenting evidence that Precambrian continental crust was present under the northwestern Pacific, and that a paleoland had existed there in the Paleozoic to Mesozoic. He based this conclusion on dredging and deep-sea drilling data, seismic profiles, paleogeography, and the geology of the Japanese Islands. However, the Chief of his Division told him that if he published the paper, he would have "no room to stay with us," while the Chief Scientist accused him of doing "bad science" and reminded him that he was on contract. Dong has been co-editor and now editor-in-chief of the New Concepts in Global Tectonics Newsletter, now in it's 10th year of publication. As the director of RAAX Australia, he essentially studies the substratum for rock discontinuities. As such, he is an expert in seismic interpretation.

To them "thanks" because they provided necessary expertise into sections of this book, and to Bruce Leybourne of GEOSTREAMS, Ltd, a geophysicist with many at-sea cruises of discovery under his belt while working for NAVOCEANO, for his understanding of the inner workings of vortices.

And to Art Meyerhoff. I will note here that this is no paean to the memory of Arthur Augustus Meyerhoff. His record speaks for itself. From the 1995 Geological Society Annual Report: he began as an economic geologist, mapping the Mendoza Province in Argentina. He worked for the USGS doing field studies in the northern Rocky Mountains. From there he landed in the energy field as an employee of Standard Oil in much of Latin America, including Cuba. He left that job to become the head publications manager for the American Association of Petroleum Geologists. He was an adjunct professor of tectonics at several universities and a guest speaker world-wide. His clients as a consultant were as diverse as Iceland, Brazil, Mexico, and the USSR. His papers are published in many languages. He could speak and read about 30 languages himself, most of which he had a passing ability to write. Truly, this man was a giant of the field and should not be ignored. AAPG called him one of the dozen eminent scientists of the 20th century. I was in his living room with him when he got the letter of acceptance from the GSA about his memoir on paleobiogeography. The letter said that they did not want to publish the data, but they could not refuse because it was all based on fact. They said it would have a great effect on plate tectonics. Art is frequently cited herein. I knew him and worked with him, and I'm proud of it.

1. GEOLOGY

Geomorphology has been defined as geology over time, acted upon by tectonics. As such, geomorphology is the key to unlocking Earth's structural history. It follows that, as a corollary, any study of Earth tectonics must of necessity address events that have and are occurring to create the ocean floor geomorphology, a body of science constituting at least 70% of Earth's surface. This treatise on that geomorphology shows the different major ocean floor feature types, their geography, bits and pieces of the accompanying geology and geophysics, and some of the processes by which they may have been constructed. The bathymetry is produced through the "eyes" of multibeam and single-beam sonar survey data. The depth of the processes is shown by the use of earthquake seismology and acoustic reflector data bases. The geology/geochronology of the processes is constrained by dredge, core, drill, and magnetics data. The tectonic processes are currently open for discussion in light of all the foregone.

Geomorphology deals with the relief that features exhibit on Earth's surface and is based on the principle that all landforms can be related to a particular geologic process or processes. A certain sequence of events explains the evolution of a feature that is dependent on time and is also predictable. The science of geomorphology is defined by the principle of uniformitarianism.

A basic understanding of the evolution and background on the derivation of the current primary working hypothesis, that of plate tectonics, is in order because tectonics is the key to unlocking the structural geometry. Lithosphere motions determine regional structure. These motions are either horizontal or vertical, such as the sinking of a basin or the rising of a mountain range. Lithosphere motion after the principal production phase gives rise to additional stresses, which may or may not change the geomorphology of an existing feature. Secondary and tertiary tectonics create secondary and tertiary features on the primary structure. Thus, the geomorphology may be used to determine the tectonic history of a particular region, or the entire Earth for that matter, and the geomorphology of discrete pieces of the ocean floor are the primary topic herein.

Historically, the Ice Ages were thought to have caused much of the surface geomorphology in Europe. That was when scientists began to drift away from the "flood" and try to decipher what they were really finding. This included the discovery of fossilized remains of dinosaur bones and plants, which were older that Bishop Ussher's 5000 or so years.

James Hutton, a Scottish geologist, was instrumental in developing the age. In an article from **Encyclopedia Britannica**: "At that time geology in any proper sense of the term did not exist. Mineralogy, however, had made considerable progress. But Hutton had conceived larger ideas than were entertained by the mineralogists of his day. He desired to trace back the origin of the various minerals and rocks, and thus to arrive at some clear understanding of the history of the earth. For many years he continued to study the subject. At last, in the spring of the year 1785, he communicated his views to the recently established Royal Society of Edinburgh in a paper entitled "Theory of the Earth, or an Investigation of the Laws Observable in the Composition, Dissolution and Restoration of Land upon the Globe." Hutton expounded that geology is not cosmogony, but must confine itself to the study of the materials of the earth; that everywhere evidence may be seen that the present rocks of the earth's surface have been in great part formed out of the waste of older rocks; that these materials having been laid down under the sea were there consolidated under great pressure, and were subsequently disrupted and upheaved by the expansive power of subterranean heat; that during these convulsions veins and masses of molten rock were injected into the rents of the dislocated strata; that every portion of the upraised land, as soon as exposed to the atmosphere, is subject to decay; and that this decay must tend to advance until the whole of the land has been worn away and laid down on the sea-floor, whence future upheavals will once more raise the consolidated sediments into new land. In some of these broad and bold generalizations Hutton was anticipated by the Italian geologists; but to him belongs the credit of having first perceived their mutual relations, and combined them in a luminous coherent theory based upon observation."

During the 1800s, another group espoused the idea of catastrophism. This hypothesis stated that most geological episodes, such as volcanism, mountain building (orogenies), earthquakes, and floods, occurred quickly and almost singularly. A worldwide event of orogeny would have necessarily been followed by a period of quiescence. The theory held that these catastrophies had decreased over time. Essentially, Earth was formed as a fireball spun off the Sun followed by an overall cooling. The outer shell cooled first, while the core remained hot. Due to the effects of gravity, the cooling outer crust collapsed in on the mantle. Unfortunately for the "catastrophists" of old, they linked their theory with the religious dogmas of the day and declared Ussher to be right; Earth was no more than a few thousand years old.

Next came Charles Lyell, another Scots geologist. In 1828 he explored the volcanic region of the Auvergne, then went to Mount Etna to gather supporting evidence for a theory of geology he was developing, that of uniformitarianism. In essence, given sufficient time, millions of years, geological change was slow and gradual and not subject to inexplicable catastrophe such as Noah's Flood. In 1829, volume one of his great work **Principles of Geology** appeared.

The geosyncline hypothesis was first developed by the American geologists James Hall and James Dwight Dana in 1859 during the classic studies of the Appalachian Mountains. Dana was first to use the term **geosynclinal** in reference to a gradually deepening and filling basin using uniformitarianism in conjunction with the concept of crustal contraction due to cooling Earth. The geosynclinal theory was further developed in the late 19th and early 20th centuries and at that time was widely accepted as an explanation for the origin of most mountain ranges until its replacement by the plate tectonic hypothesis.

From Wikipedia: "A geosyncline is a subsiding linear trough that was caused by the accumulation of sedimentary rock strata deposited in a basin and subsequently compressed, deformed, and uplifted into a mountain range, with attendant volcanism and plutonism. The filling of a geosyncline with tons of sediment is accompanied in the late stages of deposition by folding, crumpling, and faulting of the deposits. Intrusion of crystalline igneous rock and regional uplift along the axis of the trough generally complete the history of a particular geosyncline. It is then transformed into a belt of folded mountains. Thick volcanic sequences, together with graywackes (sandstones rich in rock fragments with a muddy matrix), cherts, and various sediments reflecting deepwater deposition or processes, are deposited in eugeosynclines, the outer deepwater segment of geosynclines.

"Geosynclines are divided into miogeosynclines and eugeosynclines, depending on the types of discernable rock strata of the mountain system. A miogeosyncline develops along a continental margin on continental crust and is composed of sediments with limestones, sandstones and shales. The occurrences of limestones and well-sorted quartzose sandstones indicate a shallow-water formation, and such rocks form in the inner segment of a geosyncline. The eugeosynclines consist of different sequences of lithologies more typical of deep marine environments. Eugeosynclinal rocks include thick sequences of greywackes, cherts, slates, tuffs and submarine lavas. The eugeosynclinal deposits are typically more deformed, metamorphosed, and intruded by small-to-large igneous plutons. The eugeosynclines often contain exotic flysch and mélange sediments."

From these beginnings, geosynclines, anticlines, and miogeosynclines found their way into the earth scientist's vocabulary. Tectonism is characterized by orogenesis or geosynclinal activity. Vertical tectonism became the order of the day, with erosion wearing down mountains, giant sedimentary basins forming, and isostatic compensation lifting mountains up again through thrust faulting. This could be described as "pulsation without representation" on a global scale. The rising and falling of the land allowed for the incursion by the sea, so that we had shallow oceans at one time over the mid-western United States, as an example, where the desert and high plains now reign supreme. That is why we can find fish skeletons in Wyoming. This was all very well understood by the early 1900s. Rather than merely pulsating and maintaining the same size, by the 1930s some were suspecting that Earth was actually expanding; that is, Mother Earth is getting a middle-age spread.

However, these explanations did not explain earthquakes, volcanoes, deep ocean trenches, and the apparent fit of some of the continents. Something was missing, and the geologists could not quite pinpoint the solution.

In what was then not realized as the beginning of our "modern" tectonic concept, Alfred Wegener (1915), a meteorologist, took the first step by deriving the continental drift hypothesis in his landmark book, **The Origin of Continents and Oceans**. He noticed that the eastern South American and western African continental margins seemed to fit together, so he arbitrarily placed them in close juxtaposition in the distant past and derived a new hypothesis. This idea was relegated to the "crackpot bin," where it lay for many years before being incorporated into a later hypothesis. Kiyoo Wadati (1927) first plotted the deep earthquakes around Japan. With the beginnings of earthquake seismology, B. Gutenberg and C.F. Richter predicted that Earth is a series of plates separated by active seismic belts (1949). M.L. Hill and T.W. Diblee, Jr. noted large horizontal displacements on Earth's surface along great faults (1953). Hugo Benioff deciphered the deep earthquake zones (about 650 km) around the Pacific as being interpreted as great thrust faults (1949). Bill Menard discovered (1955) the large fracture zones in the North Pacific. Ron Mason verified fracture zones by the extensive magnetic patterns on the ocean floor that ended abruptly with his work on the Pioneer in 1955. Soon after this, Keith Runcorn (1956) used magnetic pole displacement based on 180 million-year-old (Ma) rock samples to show that North America had been displaced from Europe, and this figure became the age of the oldest seafloor extant. The great rift system of the midocean ridges showed the dynamic state of the ocean floor by Bruce Heezen in 1960. Harry Hess theorized seafloor spreading in his paper, **History of Ocean Basins,** in 1962. Alan Cox, R.R. Doell, and Brent Dalrymple (1963) developed a paleomagnetic time scale using a mass spectrometer. Fred Vine and Drummond Matthews (1963) expanded on the earlier studies of Mason's magnetic anomalies. Morley came up with the same idea at that time, but he is only lately being given a share of the credit. Tuzo Wilson (1965) showed that the magnetic anomalies were offset on formation along transform faults instead of after the magnetic signature had been imprinted.

The transform faults proved to be the key, and the tectonic revolution was underway. At the 1966 Geological Society of America meeting in San Francisco, Lynn Sykes proved Wilson's hypothesis by studying earthquake motion, and Fred Vine tied all of the preceding together. The deep earthquakes and great thrust faults became the descending plate at the subduction zones. The remelt of the leading edge of that plate rose through the lithosphere in liquid (magma) form to pour out on the surface and become lava for volcanoes. The remelted lithosphere (magma) flowed on a great conveyor belt back to the midocean ridges, and there it rose to the surface to create new lithosphere. This point is recognized as the start of the plate-tectonic revolution.

Generally, the meeting of the minds revealed a fairly simple conveyor-belt explanation, and this is the gist of horizontal tectonics. Midocean ridges, or spreading centers, are zones of shallow seismicity where magma wells up to the surface from the asthenosphere, forms new oceanic crust, and moves off-ridge, all the while cooling and subsiding. This makes the spreading center a divergent plate boundary. The rate of movement varies but is measured in centimeters per year (cm/yr). The crust moves and ages with distance away from the midocean ridge crest. This idea is part of the plate hypothesis and is called "ridge push." That crust eventually, about 180 Ma, is subducted into the upper mantle in a region known variously as the subduction zone, the Wadati-Benioff zone, or the convergence zone or margin. The subduction zone is characterized by deep earthquakes and is also a convergent plate boundary. Continental and oceanic volcanic arcs form landward to Wadati-Benioff zones and are the foci of extremely virulent volcano and earthquake activity. The Pacific "ring-of-fire" is an excellent example of this phenomenon. In the vernacular, this is called "slab pull" and is one of the proposed driving forces of plate tectonics. Transform faults and fracture zones interconnect the midocean ridges and subduction zones. They are the foci of shallow earthquakes and are sometimes thought to be plate boundaries. Features not associated with those means of crustal production are generally attributed to hot spots, which Jason Morgan (1972) hypothesized to be fixed diapirs centered in the mantle. Through time, the ridges, transforms, and trenches may move about in no organized manner producing such phenomena as plate reorganizations, changes in plate movement direction, polar wandering, magnetic shifts, trench migration, and others.

Hailed by the Earth tectonic community as the universal panacea, the plate-tectonic hypothesis was adopted **in toto**. The new hypothesis answered many questions about the origins and functions of the midocean ridges, continental rifts, fracture zones, volcano production and active arcs, deep sea trenches and the attendant Benioff zones, ophiolites, accreted melanges, etc. that had never had a satisfactory explanation. And, the ocean floor was generally only known as this diagram (Figure 1).

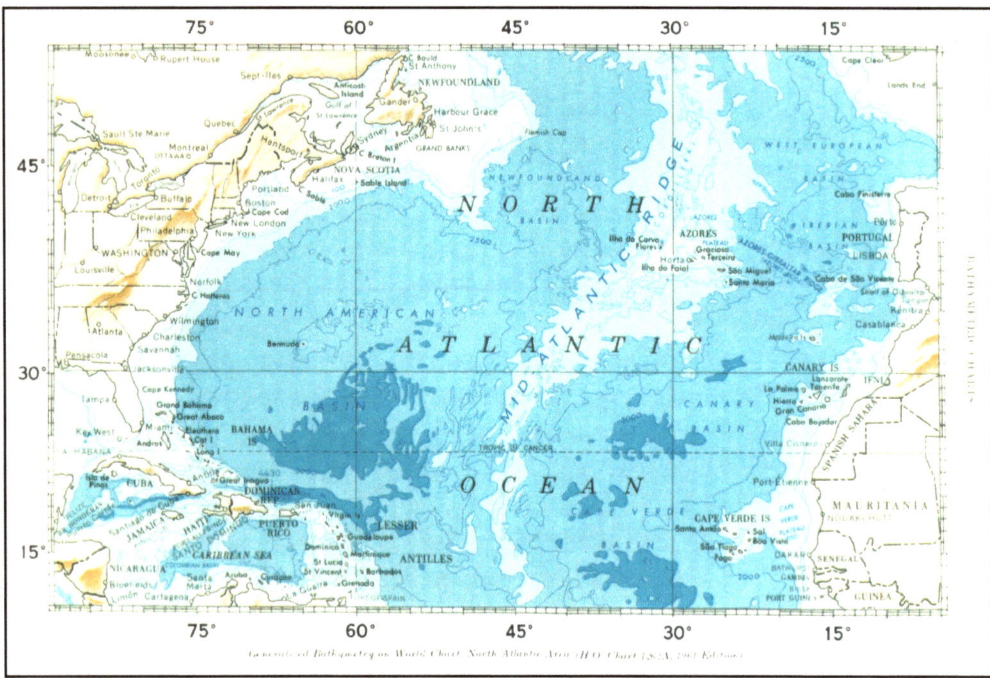

Although they are tainted by many reinterpretations, the original ideas about wandering continents were included, now called the "Bullard fit." The drifting of continents was proven by, among other things, the existence of a Triassic tetrapod fossil called **Lystrosaurus**. **Lystrosaurus** had been found in India, South Africa, and Antarctica, thus allowing the juxtaposition of all of those continents during the lifetime of **Lystrosaurus**. That cluster of continents including South America was called Gondwanaland, a super-continent (Figure 2).

Figure 1. 1961 edition of H.O. Chart 262A, the North Atlantic area.

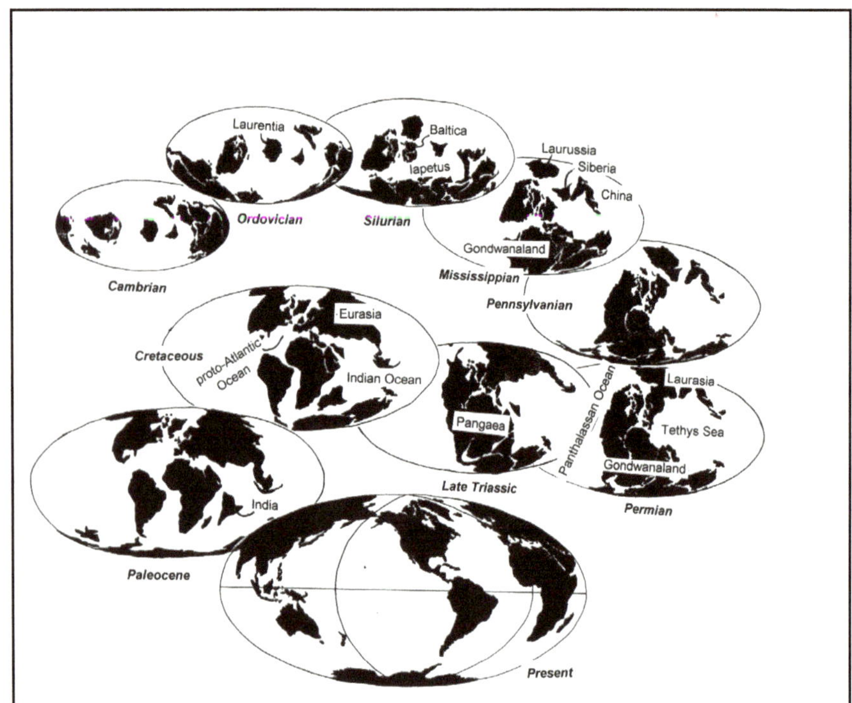

Lystrosaurus was used to prove continental drift because these continents, while being joined during the Triassic Era (about 225 million years ago (225 Ma), are no longer joined. Within the plate-tectonic continental configuration, another super-continent existed to the north, Laurasia. The extreme edges of Gondwanaland and Laurasia were in the cold oceans, and the interiors bordered the warm Tethyan Sea. Naturally, no **Lystrosaurus** could be found on both because it could

Figure 2. Continental drift and seafloor spreading for the past 600 Ma, defined by a group headed by Chris Scotese. By 200 Ma the continent called Pangaea had started to break apart, drifting towards today's configuration.

not get across such a wide body of water or such a climate extreme as must be crossed under the constraints of that continental configuration.

How did the continents wander? you may rightly ask. As noted, Earth was predicted to be a series of lithospheric plates that were separated by active seismic belts very early in the game. In the oceans, the crust is about seven km thick, while the lithosphere, which includes the crust and upper mantle, is about 80 km thick. The continental cratons are about 600 km thick, and they ride around on the oceanic plates. The plate boundaries are classified as divergent and convergent. The divergent boundaries are the midocean ridges. The convergent boundaries are subduction or collision zones where plate materials either are consumed, accreted, or built upwardly. Newer fabric may be created there as the compression forces buckle the meeting plates somewhat. Finally, the plates can slide past each other in a strike-slip action, leaving parallel-ridge and trough structures. The plates join, or weld, at different times to other plate segments, which means that they react differently at different times. The rigid mass of lithosphere, or plate, moves over a spherical surface, Earth. Of necessity, some form of random pressure must release both internally and externally to this rock. That release is in the form of fracturing, expressed bathymetrically as fracture zones by proper name. The fracture zones are Earth's great cooling cracks, and they point in the direction of plate motion. In a hypothesis that employs the principles of a conveyor belt system, everything that happens while the belt is moving affects the geomorphology.

As of this writing 12 larger plates are recognized (Figure 3): North American, South American, Pacific, Eurasian, African, Indian-Australian, Philippine, Antarctic, Caribbean, Scotia, Cocos, and Nazca. The smaller micro-plates in the plate-tectonics hypothesis are called the Mariana, Adriatic, Arabian, Aegean, Juan de Fuca, Bismark, Solomon, Fiji, Magellan, Manihiki,

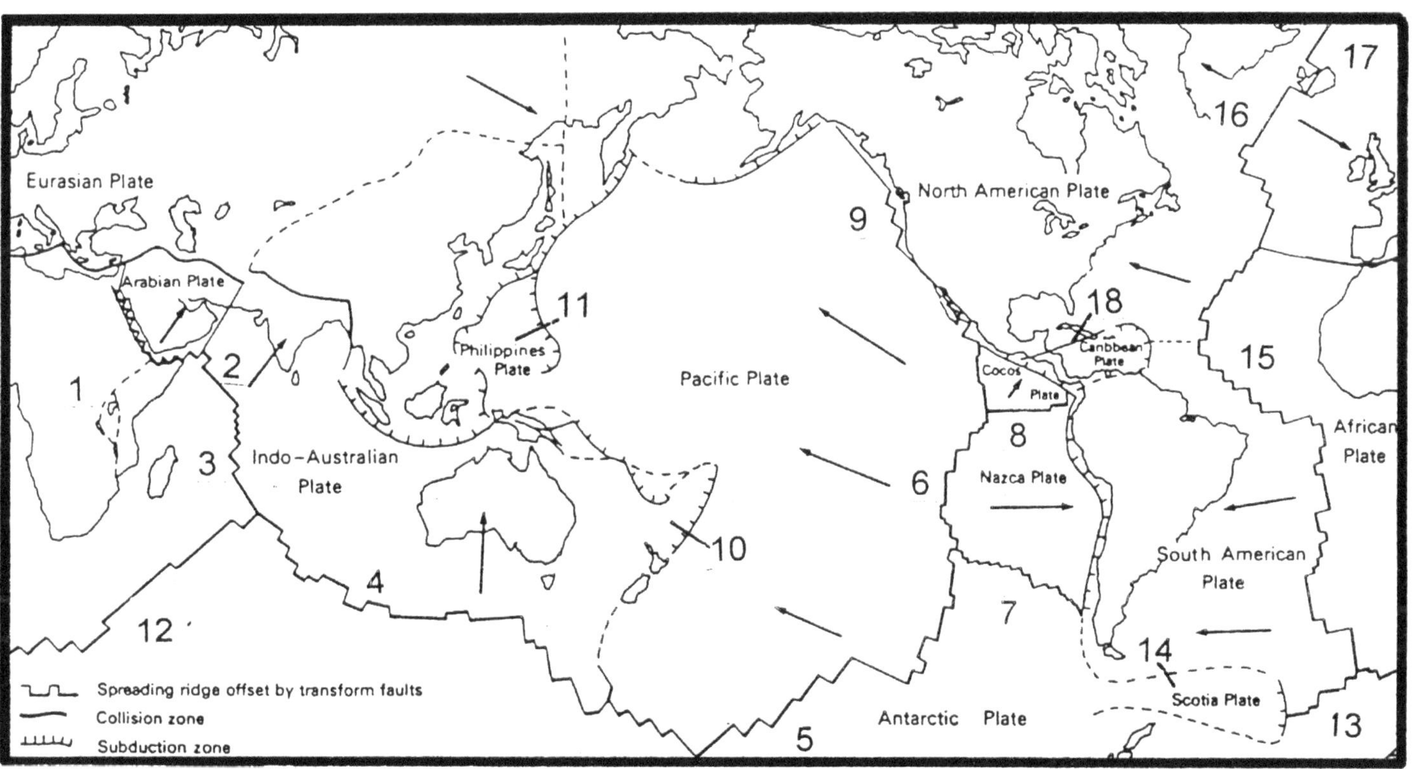

Figure 3. Plates defined by divergent (MOR) and convergent (subduction zone and collision margin) boundaries where: 1=African Rift, 2=Carlsberg Ridge, 3=Mid-Indian Ridge, 4=Southeast Indian Ridge, 5=Pacific-Antarctic Ridge, 6=East Pacific Rise, 7=Chile Rise, 8=Galapagos/Carnegie Ridge, 9=Gorda, Endeavor, Juan de Fuca ridges, 10=Lau-Havre Ridge, 11=Mariana Trough, 12=Southwest Indian Ridge, 13=Atlantic-Indian Ridge, 14=Scotia Arc, 15=Mid-Atlantic Ridge, 16=Reykjanes Ridge, 17=Mohns Ridge, 18=Cayman Trough.

Gorda, and several others. A movement is afoot to increase this number of plates almost daily as the geophysicists give us more and newer interpretations, the latest being the Indian-Australian plate split.

Two types of crust are thought to exist, continental and oceanic. Oceanic crust consists of basalts, which are primarily silicon dioxide (like sand), a fine-grained, iron-rich mafic rock. Continental crust consists of granites, which are lighter than basalt. Granite is a coarse grained, iron-poor felsic rock. Both are igneous rocks that form under different regimes.

Primary morphologic alterations followed by secondary alterations mean that the feature's history is very dynamic, and its appearance is very ephemeral on a geologic time scale. The secondary processes are related to sedimentation, both erosional and depositional. These forces have a tendency to flatten and broaden the bathymetry over time. Examples of sources of sedimentation include terrigenous, volcanigenic, eolian, biogenic, authigenic, and extraterrestrial. Marine "snow" creates its own environments. The agents of delivery to the environment include turbidity currents, volcanoes, wind, bottom currents, mass wasting, earthquakes, and **in situ** marine creatures. Post depositional processes include biogeochemical, physical/chemical, and mechanical means, such as faulting, rifting, caldera collapse, scour, landslides and slumps, sedimentation rates and types, lagoonal sediment-infilling, and subduction. In many regions the secondary effects of deposition are negated by such outside influence as current flow, which would have a tendency to "sweep" the surface clean of depositional products.

In this treatise, many of the older hypotheses will be cited, especially if they had some appropriate pieces of what will eventually be the final solution to the puzzle of Earth tectonics. However, any new hypotheses now have to have their foundation in real data. Real data have precluded the need for impromptu, **ad hoc**, and "off-the-wall" hypotheses based on mathematical models using inputs that have never been seen in the real world tied together by dreams.

Charles Wynn and Arthur Wiggins used Sidney Harris' cartoons to apprise us of their most significant scientific breakthroughs in **The Five Biggest Ideas in Science**, a 1997 book: (1) the model of the atom presented by physics, (2) chemistry's periodic table, which sorts the elements, (3) astronomy's Big Bang theory, concerning the origination of the universe, (4) the plate tectonics model of geology and geophysics to explain Earth evolution, and (5) Darwin's theory of evolution, a biology adjunct used to explain the living world. This list seems fairly comprehensive.

In this book the investigators will be named with the appropriate dates of any publications listed in parentheses. The affiliations of the investigators and degree of expertise will be noted, as will one or more of their working groups. This format is done in the interest of space, as a lengthy bibliography has been precluded in favor of more diagrams. As has been frequently noted, "one picture is worth a thousand words."

TOOLS OF THE TRADE

1. Geoglossary

So that everyone uses the same terminology, some of the working definitions are included. To begin with, the United States Board on Geographic Names (USBGN) was created by Public Law in 1947 to standardize and publish names of geographic features on an international level. Among other things, the USBGN has defined generic terms; i.e., ridges, seamounts, etc., which do not give any indication of formation processes. The terms are based merely on bathymetric contours. Many of the following definitions are taken verbatim from the USBGN, and they are under consideration for acceptance by the United Nations and the International Hydrographic Organization/ Intergovernmental Oceanographic Commission (IHO/IOC).

Marine geologists, surveyors, geophysicists, and tectonicists, or the entire marine community, have other commonly used terms concerning processes and time whose meanings are taken from the literature. All are ideal for a study of ocean floor geomorphology:

absolute plate motion - the direction and rate of motion between a lithospheric plate and the deep interior of the earth

active margin - **geosynclines**, or **convergent**

margins, these are mobile belts that are the precursors of mountains, the leading edge of the continents

anticline - Geologic structure in which the rock layers have been formed into an arch. Erosion exposes the oldest rocks at the axis of an anticline.

apron - a gentle slope, with a generally smooth surface, particularly found around groups of islands and seamounts

asthenosphere - zone of magma lying between 60 and 220-km deep between the **lithosphere** and the **strictosphere**. The P-wave velocity for this layer is 7.85-8.05 km/sec.

backarc basin - region between extinct and active volcanic arcs; a **eugeosyncline**

bank - an elevation, typically located on a shelf, over which the depth of water is relatively shallow but sufficient for safe surface navigation

basin - a depression more or less equidimensional in plan and of variable extent

bench - a small terrace

Benioff zone - earthquake hypocenters which outline an active margin between two different pieces of lithosphere

borderland - a region adjacent to a continent, normally occupied by a bordering shelf, that is highly irregular with depths well in excess of those typical of a shelf

canyon - a relatively narrow, deep depression with steep sides, the bottom of which generally has a continuous slope

collision margin - where compression raises parallel ridges and mountain ranges

continental margin sedimentary prism - a miogeoclinal prism which builds outward from the continental interior

continental rise - a gentle slope rising from oceanic depths towards the foot of a continental slope

convergent margin - the site of the juncture of the leading edges of more than one segment of lithosphere; a collision margin

cordillera - an entire mountain system including all the subordinate ranges, interior plateaus, and basins

crust - hard, very thin layer separated from the very thick **mantle** by the Mohorovicic Discontinuity (Moho); a seismic depth boundary which is only about 5 km thick under the oceans

dike - a tabular body of igneous rock that has been intruded vertically into a fissure. When erosion occurs, the remains appear as levees.

divergent margin - the site of the trailing edges of more than one plate; a spreading center

earthquake - tremor of Earth's surface caused by release of underground stress along fault lines

epicenter - point on Earth's surface directly above the earthquake

escarpment (or **scarp**) - an elongated and comparatively steep slope separating flat or gently sloping areas

fan - a relatively smooth feature normally sloping away from the lower termination of a canyon or canyon system

fault - A fracture or break in Earth's lithosphere along which differential movement of the rock masses has occurred.

flank break depth - approximate measurement where relatively steep upper slopes change to gentler lower slopes around the original edifice, generally due to excess sedimentation

flank rift zone - spinelike protuberances extending from the centers of volcanic edifices; a **dike**

flat - a small level or nearly level area

foot wall - The side of the fault containing the rock layers that are below the fault plane.

forearc - region between active volcanic arc and trench

fork - a branch of a canyon or valley

fracture zone - an intensive linear zone of irregular topography of the ocean floor characterized by steep-sided or asymmetrical ridges, troughs, or escarpments (the ages of the ocean floor on opposite sides of the fracture are usually different). A fracture zone may also be called a **fracture swarm** when it splays, a **hot line** if leaky, or a strike-slip fault; a compression crack.

furrow - a closed, linear, narrow, shallow

depression

gap - a narrow break in a ridge or rise

geanticline - area from which sediments are derived; opposite of **geosyncline**

geodynamics - a term generally used by geophysicists to describe the soft, slow processes within the Earth's surface, almost entirely driven by heat transfer

geosyncline - a downwarped portion of the Earth's crust; a very large basin in which many thousands of feet of sediment accumulate; a trench, its sediment fill, and its forearc wedge.

gully - a small valley-like feature

guyot - a seamount having a comparatively smooth, flat top which has undergone some form(s) of subaerial erosion

hanging wall - The side of the fault containing the rock layers that are above the fault plane.

hill - an elevation rising generally less than 500 meters

hole - a small depression of the ocean floor

hypocenter - location inside Earth where earthquake originates

knoll - an elevation rising generally more than 500 meters and less than 1000 meters and of limited extent across the summit

ledge - a rocky projection or outcrop, commonly linear and near shore

levee - an embankment bordering a canyon, valley, or seachannel

lithosphere - the solid layer lying between the hydrosphere (ocean) and the **asthenosphere**, usually considered to be about 80-km deep, but with a range of from 0-180 km. It is sufficiently cold to be brittle and is considered to be the "**plate**." The P-wave velocity of this layer is 0-8.45 km/sec.

mantle - rock layer between the **crust** and the core

median valley - the axial depression of the midocean ridge system

midocean ridge - the divergent margin which is an upraised, world-encircling feature where new ocean floor is created

miogeosyncline - sometimes called an **ideogeosyncline** or a **germano-type** geosyncline, This would necessarily rise up to become mountains after folding, thrusting, metamorphism, intrusion of igneous rocks, and uplifting.

moat - an annular depression that may not be continuous, located at the base of many seamounts, islands, and other isolated elevations

mobile belt - Long, relatively narrow region of tectonic activity (past or present) that delimit cratons. Mobile belts contain a thick, mostly complete column of strata, but the rock record is hard to read because of folding, faulting, unconformities, metamorphism and igneous intrusion.

mound - a low, isolated, rounded hill

mountain - a well-delineated subdivision of a large and complex positive feature

neovolcanic zone - locus of most recent volcanism which straddles the spreading axis

normal fault - A fault in which the hanging wall has moved down relative to the foot wall.

ophiolite melange - Once called a **eugeosyncline**, this rock association is produced at the axial rift valleys of midocean ridges.

orogenic/tectogenesis phase - mountain building, where flysch is the synorogenic phase; molasse is the post-orogenic; and igneous activities can be any of the phases

overlapping spreading center - en echelon overlapping of offset **neovolcanic zones**, creating basin in between; an eddy-type of vortex structure

passive margin - Once named **miogeosynclines**, they are now considered to be a continental edge where there is no stress state and no subduction.

peak - a prominent elevation, part of a larger feature, either pointed or of very limited extent across the summit

pinnacle - a high tower or spire-shaped pillar of rock or coral, alone or cresting a summit

plain - a flat, gently sloping or nearly level region

plateau - a comparatively flat-topped feature

of considerable extent, dropping off abruptly on one or more sides

propagating rift - en echelon, paired, V-shaped pseudofaults resulting from single propagation sequence with V pointed in direction of propagation

province - a region identifiable by a group of similar physiographic features whose characteristics are markedly in contrast with surrounding areas

ramp - a gentle slope connecting areas of different elevations

range - a series of associated ridges or seamounts

ravine - a small canyon

reef - a hazard to navigation with a depth of 20 meters or less, composed of consolidated rock

relative plate motion - the direction and rate of motion between lithospheric plates

reverse fault - A fault in which the foot wall has moved down relative to the hanging wall.

ridge - a long, narrow elevation with steep sides

ridge segmentation - pinch-and-swell geometry of all mountain belts (White, 1989) where the average distance between pinches ranges from 50 to 300 km. The pinches are underlain by **transform faults** and **overlapping spreading centers**.

rise - a broad elevation that rises gently and generally smoothly from the ocean floor

saddle - a low part, resembling in shape a saddle, in a ridge or between contiguous seamounts

seachannel - a continuously sloping, elongated depression commonly found in fans or plains and customarily bordered by levees on one or two sides. The seachannel load commonly builds the **plains**.

seamount - an elevation rising generally more than 1000 meters and of limited extent across the summit

shelf - a zone adjacent to a continent (or around an island) that extends from the low water line to a depth at which there is usually a marked increase of slopes towards oceanic depths

shelf valley - a valley on the shelf, generally the shoreward extension of a canyon

shoal - a hazard to navigation with a depth of 20 meters or less, composed of unconsolidated material

sill - the low part of a gap or saddle separating basins. In geology a sill is a tabular body of igneous rock horizontally intruded between sedimentary or volcanic beds or along foliation planes of metamorphic rocks.

slope - the slope seaward from the shelf edge to the beginning of a continental rise or the point where a general reduction in slope occurs

spur - a subordinate elevation, ridge, or rise projecting outward from a larger feature

strictosphere - hard shell underlying asthenosphere having a P-wave velocity of 8.1-8.7 km/sec

strike-slip fault - A fault in which the displacement along the fault is horizontal.

structural basin - Geologic structure in which the rock layers dip in toward the center of the structure, usually in concentric circles. Erosion exposes the oldest rocks at the edges of the structure.

subduction zone - **convergent margin** where the seaward lithosphere is downbuckled and descends into the **asthenosphere**. The process of consuming **lithosphere** is subduction. The parts of a subduction zone in order are seaward **lithosphere** and **seamount**s, **trench**, inner trench wall, outer arc high, and **forearc**.

summit plateau - area inside the summit plateau break depth

summit plateau break depth - point at which eroded portion of the feature stops and the original slopes begin

swale - a low place between adjacent highs

syncline - Geologic structure in which the rock layers have been formed into a trough. Erosion exposes the oldest rocks at the edges of a syncline.

tectonic spreading fabric (**TSF**) - small lineations parallel to **midocean ridge**, originating in ocean floor formational process

tectonics - study of the deformation of the rocks that make up the Earth's crust and the forces that produce such deformation. It deals with the folding and faulting associated with mountain building; the large-scale, gradual, upward and downward movements of the crust; and sudden horizontal displacements along faults. Other phenomena studied include igneous processes and metamorphism. The chief working principle of tectonics is the concept of plate tectonics.

tectonostratigraphic terrane - exotic terrane, or body of rock alien to the country rock; **dike**s and **sill**s

terrace - a relatively flat horizontal or gently inclined surface, sometimes long and narrow, which is bounded by a steeper ascending slope on one side and by a steeper descending slope on the opposite side

thalweg - an imaginary line connecting the low points of a canyon floor

tongue - an elongate (tongue-like) extension of flat ocean floor into an adjacent higher feature

transform fault - perpendicular offset on spreading axis between ridge segments (Wilson, 1965b). This is the active portion of a **fracture zone**; also a **ridge-transverse fault zone**.

trench - a long, narrow, characteristically very deep and asymmetrical depression of the ocean floor, with relatively steep sides. This is where the warmer landward **lithosphere** overrides the colder oceanic **lithosphere**. The region immediately beneath the trench is the zone of engulfment.

trough - a long depression of the sea floor characteristically flat bottomed and steep sided, and normally shallower than a **trench**

valley - a relatively shallow, wide depression, the bottom of which usually has a continuous gradient (This term is generally not used for features that have canyon-like characteristics for a significant portion of their extent.)

volcanic arc - extruded magma in volcanic form at **active margin** on overthrusting landward lithosphere

vortex structure - large ovate, oval to circular, area 200--1000-km in diameter features which are probable areas of either mantle diapirism (upwelling) or downwelling

In a philosophical discussion related to several of the definitions above, we are immediately presented with differences of opinion. The crust is at once listed as occupying many layers. It is shown to be the lithosphere and the asthenosphere. The 15-km thick oceanic crust is comprised of sediments, basaltic pillow lava, a doleritic intrusive complex, and a gabbro-basaltic layer before reaching the Mohorovicic Discontinuity (Moho for short). Another interpretation lists the 5-km thick oceanic crust as the outer shell of the 100-km thick lithosphere, separated by the Moho. From yet another standpoint, the divisions are as plates. Plates are comprised of the oceanic crust, which is the upper 10-30% of the lithosphere, and the lithosphere itself, which is about 100-km thick. The lithosphere also consists of a thick slice of the mantle. The plate is rigid and floats on the asthenosphere. For our purposes, we will adhere to the above listed definitions and leave the philosophy to those who are more qualified to lead such a discussion. The crust is the top, hard layer. The lithosphere is next, and it rides on the asthenosphere.

2. Bathymetry

Morphologically, very few ocean floor features were correctly portrayed in the literature before the 1980's. Until that time there was essentially no science of intermediate- to small-range geomorphology. Regions, provinces, and large features were known. The Indian Ocean is a case in point. The midocean ridges and gross fracture valleys are readily apparent but may not remain valid under close scrutiny with a multibeam system. The Germans produced an accurate rendition of Great Meteor Tablemount in the northeast Atlantic by concentrating on surveying the feature using a single-beam sonar. Because it looks like the results of a systematic survey, the Great Meteor Tablemount is the first feature to be portrayed accurately. All the rest were merely exploratory lines or an occasional star pattern. Any accurate morphology was purely coincidental. The first multibeam representation of seamounts was published in the 1970s. Since that time, enough multibeam data are available to begin a study of the ocean floor.

Interestingly, the multibeam sonar system and the plate-tectonic hypothesis evolved at about the same time and have since been co-evolving, albeit in diametrically opposed directions. However, the hypothesis was in place first before any meaningful ocean floor surveying was completed, so we bow to the elder statesman in the early going of our geomorphologic explanations.

In order to have a meaningful study of the geomorphology of the ocean floor, one must be able to "see" that floor. This is by means of bathymetry. The multibeam Sonar Array Sounding System (SASS) was put into operation on 12 July 1967 by the US Navy's Naval Oceanographic Office (NAVOCEANO). To collect multibeam sonar bathymetry, a burst of acoustic energy is transmitted from the sonar projectors on the ship's hull. Time delays between the highest and lowest projectors are pre-programmed with roll and pitch compensators, and a fan-of-sound is sent out. The series of 1°X1° beam width hydrophone receivers converts the returning acoustic energy to electrical signals that provide input into the receivers. The return is processed into range and angle data, which is translated into raw depth and location. With the sound-velocity profile having been entered into the program, ray-bending predictions are made and applied. The output is now in corrected depths and lateral positions.

The swath mappers, utilized in organized survey patterns, were/are able to provide us with a sonar-bottom map on any ocean floor region desired. The sonar-bottom is defined as that set of contours produced using a distinct sound velocity profile in the compilation program. Thus, the surveyor can set a constant value, such as 4800 ft/sec, and anyone else can recover on that same set of contours using the same sound velocity. One can already see that the setting of this value is probably the single most important factor in collecting any sonar data, as the speed-of-sound in seawater is used to determine the two-way travel time, and ultimately the depth value.

The US Navy was the first to use such a system. Guessing about actual features became passe because of the sonar collection process employing a fan-of-sound and updated sonar equations. That fact alone established the SASS as the world's premier sonar collector, and NAVOCEANO was the only collection agency to use the SASS system. The data were collected in fathoms (fm) and nautical miles (nmi) because that was the standard unit of measurement for the submarine community. For the purposes of this treatise, one fathom equals 1.83 meters equals six feet. One degree of latitude equals 60 nautical miles equals approximately 100 kilometers.

In a form of technology transfer, the makers of the SASS system (General Dynamics) were allowed a bit of free enterprise, and they produced a less-accurate system for consumption by the paying public. That system was called the SeaBeam. While the data collected are not as refined, they are certainly adequate to describe ocean floor features.

Additionally, during the 1970s the World Chart Group at NAVOCEANO was tasked with collecting all the world's ship-of-opportunity single-beam sonar data. They were responsible for producing the world's first gridded bathymetric data base, called the DBDB-5. It is a five-minute grid, meaning that it uses one data point every five miles (Figures 12,13, 15, 33, and 51).

The caveat is raised that using a finer beam width sonar does not necessarily show all types of features, especially in the modern era of computer cartography. The computer grid pattern will not show steeper features such as scarps, outcrops, gullies, and holes. Nevertheless, remotely operated vehicles (ROVs) are showing that they do exist but do not appear on the multibeam strip. Contouring techniques, such as the proverbial "cartographic license," have also done much to prevent proper geomorphology, especially where longer contours have been foreshortened excessively to create "dream features." Many previously defined features could have been combined. The best example of this is the Mapmaker Seamounts in the north-central Pacific, which are for the most part the long, linear parallel ridges of the Mendocino Fracture Zone. Previous worldwide charts have incorporated many misconceptions as to the frequency and geomorphology of the ocean floor.

Since 1966 many updated maps of the ocean floors have become available. Another option is to draw upon all the survey information gathered since that time from NGDC. NGDC is the repository for all bathymetry data collected by ships-of-opportunity, especially those operating under US National Science Foundation (NSF) grants. Those investigators are required to send their survey data to the NGDC. Therefore, one can purchase updated bathymetry CDs on a regular basis, filter the data to suit oneself, and become a nautical cartographer. This will preclude any misunderstandings of the actual bathymetry in many cases. Obviously, the entire ocean floor has not been surveyed, but enough exists in many regions to do a very respectable job of updating the original 1970s maps.

A three-dimensional GIS package, called GRASS, was developed for NAVOCEANO in the 1990s. Any size pixels could be used as long as the data supported the pixel size. Essentially, evenly spaced grids are obtained using a multi-stage two-dimensional interpolation algorithm (CHRTR). This interpolator was specifically designed to permit a faithful reproduction of the input data in the data dense areas and to transition gracefully to a regional field of sparse data. It incorporates anti-aliasing filters, a minimum curvature spline algorithm to generate the regional grid, and a merge algorithm to blend input and region data to produce the final output grid.

In the case of many of the 3Ds shown herein, that size is on the order of one-tenth of a minute/ nautical mile with no sparse data fields. Where the diagrams cover too large a territory, that limit is expanded accordingly, such as a 0.5-minute grid. Considering the fact that many of the seamounts cover over 4800 square miles, reduced to an 6"X6" diagram, this is very close order information. The larger the grid size, the less will be the regional character.

3. Geochronology and Geology

In a study of geomorphology, one of the integral parts of the equation is the timing of the events. This has historically been done in two ways, by actual rock dating and by the geophysical surveying technique producing magnetic anomalies. Because the original assignment of values to the magnetic stripes was made by using actual rock ages, one could assume that they were exact, or nearly so, ages of the ocean floor. This said, the oldest floor was then shown to be on the order of 200-Ma and everyone got on board with that date.

While the study of magnetic anomalies is not really germane, that topic is discussed at this time. Magnetic anomalies are based on reversals of the magnetic field at discrete intervals. This was first hypothesized by Bernard Bruhnes about 95 years ago when he noticed reversed polarity of clays associated with lava fields. In the 1950s Seiya Uyeda and Takesi Nagata from Tokyo University demonstrated that certain rocks acquired a magnetization just the opposite of the field where they cooled. The fact that rocks younger than 780-Ka are all magnetized in the same direction was discovered by a host of studies in 1963. This work was verified by Christopher Harrison and Brian Funnell at Scripps Institution of Oceanography (SIO) by sediment cores from the Pacific Ocean floor. Morley, Vine, and Matthews combined the reversals with Hess' idea of seafloor spreading. Since that time, about 1966, this measurement has been used extensively to date the ocean floor while proving the plate tectonic hypothesis.

Field hands are not satisfied with the internal audit provided by the hypothesis, most preferring to work with actual samples; that is, rocks. Very early in the study of plate tectonics, some higher authority determined that the anomalies needed to be ground truthed. Rock samples were determined to be the source of that truth; that is, samples collected at the source. This was certainly a viable and worthwhile undertaking. The DSDP was formed in 1968, two years after the formulation of the working hypothesis. The purpose of this project was to obtain basement material at predetermined sites which would solidify the magnetic anomaly values, and this is what they discovered:

The GLOMAR Challenger was placed into operation, NSF was instrumental in selecting the principal investigators, and the process of proving the ocean floor age progressed. In a study of the first 429 sites by Art Meyerhoff, not one off-ridge drilling attempt reached basement. In a continuation of that study, DSDP Sites 430 through 625 and ODP Sites 626-949 have been investigated for an additional 520 sites. The categories are divided into (1) on-ridge basalts (pillow, fractured, breccia, or enriched), backarc basins, and emplaced large igneous provinces; (2) off-ridge, seafloor basalts; and (3) basalt not reached. The first yielded 111 sites. The midocean ridges, several continental borders, and the Lau Basin and the Philippine Sea were all Miocene or younger. Sites 443 and 444 are in the Shikoku Basin, and the rock ages disagree with the magnetic ages. The various rises, banks, atolls, guyots, and plateaus revealed a variety of ages, all emplaced features on an older ocean floor. Those were listed as Eocene to Oligocene in age. It is the two other categories that yield ages that are germane and are not in line with the original intentions of the DSDP. Site 801 in the NW Pacific was found to be 157 Ma interbedded basalts with some pillow basalts, which is not basement. No basalt was recovered from the other 395 cores.

This study demonstrates that, while we know much about the overlying sediments on or near the large, igneous provinces, we still know very little about the age of the off-ridge basement or it's composition. Too many of the descriptions in those volumes read "sills or dykes, fractured basalt interfaced or interbedded with other sediments, andesitic mixed with basalts" and others. These descriptions do not inspire a great deal of confidence in the would-be user. Apparently at the end of the DSDP basement could not be reached. The formation of chert at depth is suspected to be the "ceiling" through which the drills cannot penetrate. The practice was stopped by the ODP. Therefore, the age of the magnetic anomalies is not ground-truthed and will not be included in this discussion.

The Ocean Drilling Program (ODP) is funded by the U.S. National Science Foundation and 22 international partners (JOIDES) to conduct basic research into the history of the ocean basins and the overall nature of the crust beneath the ocean floor using the scientific drill ship JOIDES Resolution. Joint Oceanographic Institutions, Inc. (JOI), a group of 18 U.S. institutions, is the Program Manager. Texas A&M University, College of Geosciences is the Science Operator. Columbia University, Lamont-Doherty Earth Observatory provides Logging Services and administers the Site Survey Data Bank.

Any opinions, findings, and conclusions or recommendations expressed in these documents are those of the author(s) and do not necessarily reflect the views of the National Science Foundation, the participating agencies, Joint Oceanographic Institutions, Inc., Texas A&M University, or Texas A&M Research Foundation.

All the while this mass wasting of taxpayer dollars has been going on, scientists not funded by the NSF have also been taking ocean floor samples by whatever means. What they have recovered from the ocean floor has been compiled by various countries into newer, more updated ocean floor chronology maps. They look nothing like the original age maps based on magnetics. In fact, they are unrecognizable in a plate tectonics setting. Be that as it may, the real rock ages and types are introduced throughout this book at the appropriate times. You will be able to draw your own conclusions; I do not need to drive them.

Oceanic lithosphere is a 10-km thick and generally uniform shell, which is mostly basalt. As a last resort, earth scientists have used the magnetic stripes found ubiquitously. The older stripes are called the "M" series. The ages listed for the "M" series are not ground-truthed, so that measurement is merely used to show a chronology in relative ages, as in "this" happened before "that." Additionally, the current magnetic maps of the ocean basins include the catch-all, non-descriptive terms "Cretaceous Magnetic Quiet Zone" and "Jurassic Magnetic Quiet Zone." This is because nobody had the data to make the determination as to the ages of the ocean floor in those regions.

It is interesting to note a quote by the then-current head of the ODP, Tom Davies (1998): "... indeed, we have yet to find rocks older than 200 MY in the deep oceans. These observations lead to a rapid acceptance of the plate tectonic hypothesis." Obviously, this was a mistake on his part.

A time scale is added in Appendix A.

4. Geophysics

A. Earthquake Seismology

Earthquakes played an integral role in the derivation of the plate tectonic hypothesis by explaining and outlining several facets that had been previously unexplained. As noted, tectonics is the key to unlocking the structural geometry, and the regional geology is the key to unlocking the tectonic history. Lithosphere motions determine regional structure and the geomorphology. Lithosphere motion after the principal production phase gives rise to additional stresses which may or may not change the geomorphology of an existing feature. The direction of lithosphere motion is imprinted by the fracture zones and megatrends. A basic understanding of the earthquake seismology associated with the lithospheric motion, and a basic understanding of what the fracture zone show, is essential to understand Earth tectonics.

Earthquake seismology first entered into the tectonic explanations when Earth was predicted to be a series of plates separated by active seismic belts by Gutenberg and Richter (Figure 3). Large horizontal displacements on Earth's surface along great faults, or fracture zones, were noted. Benioff interpreted the deep earthquake zones around the Pacific to be great thrust faults at the surface, or vertical displacement. The lower earthquakes were noted to be "decoupled" from the upper; that is, they did appear on the same strike or even to be part of the same tectonic regime. Last, shallow earthquake zones underlying the midocean ridges were noted early on, and they were used to help show where the central meridian of the ridge could be in the early bathymetry efforts. Thus, four different sets of earthquake parameters were noticed for tectonic events on the ocean floor.

Because the earthquake data were originally used to define the active margin, they are included for most of the world. Earthquake data show activity at depth in the upper mantle. Earthquake activity occurs under the midocean ridges, along fracture zones, under volcanoes, etc., but more often the activity is located at the trenches. Originally, an International Seismic System (ISS) collected the earthquake information and sent it to the NGDC in Boulder, Colorado, for distribution. That function was shifted from the ISS to the U.S. Geological Survey (USGS) in 1961. Much of the earthquake information was reported erroneously, and NGDC has updated the data base. The 1990 CD-ROM was used for the table constructions herein. The 2006 USGS data base was used to construct Appendix B. The numbers may change, but probably by not enough to make an appreciable difference. As an example, if no deep earthquakes are listed in an area, then the removal of erroneous data in that area will not introduce deep earthquakes there.

The earthquake regime is defined as: shallow earthquakes down to 70 km, intermediate earthquakes between 70 and 300 km, and deep earthquakes deeper than 300 km. Theoretically, the frequency decreases logarithmically down to 300 km, then increases again to 550 to 600 km. There are no earthquakes beyond 660 to 700 km.

Some layer the earthquake regimes at phase change discontinuities at 450 km, where olivine, a major constituent of basalt, changes to spinel, and another at 650 km, where spinel undergoes a phase change to form either perovskite or magnesiowustite. This is only where olivine still exists. M. Wysession, for instance, does not account for the first phase change to eclogite, a metamorphic, high-pressure equivalent of basalt. Eclogite contains pyroxene and garnet. Between the depths of 80 and 150 km, the upper mantle composition may be up to 50% eclogite. That is a very significant phase change to ignore. Eclogite can come to the surface in a rapid-transport setting and has been found naturally. The importance of the spinel and perovskite/magnesiowustite phase changes is theoretical and is not yet based on sampled data. No holes have been drilled to the necessary depth.

The trenches are hypothesized to be the only place where intermediate to deep earthquakes exist. Scientists have examined the idea of phase changes in the olivine to explain deeper earthquakes, but the explanations have been lacking in key areas. Seismologists and tectonicists have derived a new "proof." Faulting occurs in the descending slab, and this accounts for the shallow earthquakes. As the hydrated oceanic crust sinks it is slowly heated. It becomes dehydrated and becomes "fluid-assisted faulted." The slab's interior remains cold so that the inner olivine in the subducting slab cannot transform into spinel. At 300 km the outer slab temperature increases and enough dewatering takes place to allow anti-crack faulting as the spinel phase changes begin. This process creates intermediate earthquakes. As the slab descends deeper, the inner olivine follows suit, and earthquakes again increase to about 600 km. At that point all earthquake activity ceases.

Table I. Summary earthquake data for all active margins.

(total number of earthquakes in study from NGDC earthquake 1990 CD-ROM is 182,597)

Earthquake depths (kms)	Quantity	% of total	
0-49	121,557	66.6	
50-99	25,633	14.0	
100-149	14,481	7.9	
150-199	6941	3.8	
200-249	4008	2.2	
250-299	1357	0.7	
300-349	801	0.4	supposed increase at
350-399	856	0.5	eclogite phase change
400-449	905	0.6	
450-499	871	0.5	
500-549	1569	0.8	increase at spinel
550-599	2060	1.1	phase change
600-649	1341	0.7	
650-699	212	0.1	
700+	5	-	
unknown	5300+		

Of the more than 182,000 events on record, 70.6 % of all earthquakes appear in the region between 0-100 km. These are the shallow earthquakes. Where mushy magma exists in the asthenosphere between 100 and 300 km, 14.6% of the earthquakes appear. These are the intermediate earthquakes. Further study shows zones in this layer at discrete depths that contain no earthquake activity. Only 4.7% of the earthquakes occur below 300 km, the area where the subducting slab is thought to be melting and providing the excess magma to build the volcanic arcs. Please notice that there is no increase at the 300-km boundary.

From the US Geological Survey (USGS) Internet site: Magnitude is measured by the Richter Scale, which measures the magnitude of seismic waves from an earthquake, devised in 1935 by the American seismologist Charles F. Richter (1900-1985). The scale is logarithmic; that is, the amplitude of the waves increases by powers of ten in relation to the Richter magnitude numbers. The energy released in an earthquake can easily be approximated by an equation that includes this magnitude and the distance from the seismograph to the earthquake's epicenter. The Ricther scale is an open - no lower or upper limit - scale, but no earthquake greater than about 9 has been recorded. An earthquake whose magnitude is greater than 4.5 on this scale can cause damage to human-built structures provided they are close enough to the epicenter; severe earthquakes have magnitudes greater than 7. The famous San Francisco earthquake of 1906 was 7.8 on the Richter scale; the Alaskan earthquake of 1964 was 8.4; and the Loma Prieta quake of 1989 was 7.1. Like ripples formed when a pebble is dropped into water, earthquake waves travel outward in all directions, gradually losing energy, with the intensity of earth movement and ground damage generally decreasing at greater distances from the earthquake focus. In addition, the nature of the underlying rock or soil at a particular location affects ground movements.

In order to give a rating to the effects of an earthquake in a particular place, the Mercalli scale, developed by the Italian seismologist Giuseppe Mercalli, is often used. It measures the intensity, that is, the severity of an earthquake in terms of its effects on the structures and the inhabitants of an area, e.g., how much damage it causes to buildings and whether or not sleeping persons are awakened by it. The Mercalli scale is a closed one, ranging from 1 to 12, at which upper limit all structures collapse; total destruction.

Earthquakes with magnitude (M_w) of about 2.0 or less are usually called micro-earthquakes; they are not commonly felt by people and are generally recorded only on local seismographs. Events with magnitudes of about 4.5 or greater - there are several thousand such shocks annually - are strong enough to be recorded by sensitive seismographs all over the world. Great earthquakes, such as the 1964 Good Friday earthquake in Alaska, have magnitudes of 8.0 or higher. On the average, one earthquake of such size occurs somewhere in the world each year. The Richter Scale has no upper limit.

A prodigious amount of heat, on the hypothesized order of 35 billion megawatts, should not have happened at this depth. The pressure equivalent at this depth is about 200,000 atmospheres, so an excess amount of melting should occur. Hypothetically, this is the very bottom of the descending slab, a slab which begins the melting process many kilometers above this point as it feeds fresh magma back to the surface in the form of active arc volcanism. Where deep fault zones have been exposed on the surface, no evidence of any kind of heat or melting can be found, though.

Additionally, intraplate continental earthquakes are common; far removed from the active margins. Most of the mountain ranges are underlain by belts of earthquakes. Many major earthquakes have occurred in recent history that have been extremely vigorous and dangerous to humans. A robust working hypothesis must include an explanation for the cause, and existence, of the major continental earthquakes in recent history (NEIS). For the Indian subregion: Kangra (4 Apr 1905; mag 8.6), Bihar (15 Jan 1934; mag 8.4), Quetta Pakistan (30 Apr 1935; mag 7.5), Khillari (29 Sep 1993; mag 6.2). In Asia: Central Asia (21 Oct 1907; mag 8.1), Gansu, China (16 Dec 1920; mag 8.6), Yunan, China (16 Mar 1925; mag 7.1), Xining, China (22 May 1927; mag 8.3), Gansu, China (25 Dec 1932; mag 7.6), Erzincan, Turkey (26 Dec 1939; mag 8.0), Qazvin, Iran (1 Sep 1962; mag 7.3), Yunan Province, China (4 Jan 1970; mag 7.5), Southern Iran (10 Apr 1972; mag 7.1), Tangshan, China (27 Jul 1976; mag 8.0), NW Iran-USSR Border (24 Nov 1976; mag 7.3), Iran (16 Sep 1978; mag 7.8), Turkey-USSR Border (7 Dec 1988; mag 7.0), and Western Iran (20 Jun 1990; mag 7.7). The same holds true for Australia, a region with no spreading centers and no subduction zones: Adelaide (3 Jan 1954; mag 5.4), Robertson/Bowral (22 May 1961; mag 5.6), Meckering (14 Oct 1968; mag 6.8), Calingiri (10 Mar 1970; mag 5.5), Picton (10 Mar 1973; mag 5.5), Cadoux (2 Jun 1979; mag 6.2), Lithgow (13 Feb 1985; mag 4.0), Marryat Creek (30 Mar 1986; mag 5.8), Tennant Creek (22 Jan 1988; mag 6.8), Newcastle (22 Dec 1989; mag 5.6), and Ellalong (6 Aug 1994; mag 5.4). In the United States we found the New Madrid earthquake in 1812 (mag 7.9), 1811 (mag 7.7), and again in 1812 (mag 7.6). Nevada listed 1915 (mag 7.7) and 1954 (mag 7.3). Idaho had one in 1983 (mag 7.3). Bear in mind that no small or micro-earthquake activity is in these lists. However, if the reader would like to take the time, the NEIS webpage lists 1167 for India alone, going back to about 1000 A.D. Suffice it to say, many continental earthquakes have occurred; most have not been explained in a proper tectonic framework.

The Shikotan earthquake in Kamchatka presents another anomaly. This 1994 M_w=9.3 quake occurred in an intraplate transverse zone in the Kuril-Kamchatka arc. It again demonstrated a "non-subductional mechanism." Credit where it is due, the 1997 Kronotskoe earthquake is included. With an $M_{w=}$7.9, and an $M_{s=}$7.5, this quake defined the subduction plane as presumed.

Horizontal earthquakes also exist, as in strike-slip zones. The Landers earthquake, occurring 28 June 1992 in southern California, was a magnitude M_w 7.3 right-lateral strike-slip to the depth of 1.1 km. The average slip was about 3-4 m, with 6 m being the max. The accelerated amount of strain following this earthquake resulted in anomalous post-seismic displacement of the crust ranging from a few mm to 55 mm along the fault, resulting in the Big Bear aftershock, which registered M_s 6.4. The strain released in the upper 10 km of the lithosphere can be easily explained by aftershocks. However, the major release came from below 10 km, showing viscous relaxation along more than 100 km of the fault.

Of the several earthquakes around eastern Kamchatka, one stands out in particular. The 1995 Neftegorsk quake on Sakhalin Island, with an $M_{w=}$7.1, and an $M_{s=}$7.6, was huge. There was a rupture length of 40 km, and the fault plane is nearly vertical. There was an absence of aftershocks, but the motion was as expected by the plate-tectonic interpretation for this region.

The Mendocino Megatrend was long ago predicted to continue to at least Yellowstone National Park in NW Wyoming. We know that earthquakes are associated with megatrend activity. A report from the Utah seismic network (UUSS) indicates at least 13 events in the park from 10-17 October 2003, with an additional 10 in the Teton Range. Mapping the events from 1979 to the present gives a clear outline from NW Yellowstone through the park to the ESE, turning south, and continuing SSW through the Tetons. The present swarm has M_s ranging from 0.4-1.9, all lying above the 350° isotherm in the brittle crust. Some of the roads have been shut down to tourism, and many of the birds have left the area. This is similar to tectonic events before the eruption of Mt. St. Helens in the early 1980s. And, Yellowstone is not anywhere near a plate boundary/subduction zone.

The USGS has been studying earthquakes in New Mexico. Their data, collected between 1962-1998, shows no evidence at all of any activity along the state's major structure, the Rio Grande Rift.

Into this we need to mix the June 1994 Bolivian earthquake. This deep earthquake, an M_w=8.3 at 636-km deep, supposedly extended across a 30 by 50 km plane. It cut horizontally across the slab and extended well beyond the supposed olivine layer. The significance of this fact is that, heretofore, the motion of earthquakes at depth had been expected to be nearly vertical. The horizontal movement of the Bolivian earthquake is difficult to explain within the current constraints. Big, deep earthquakes should not occur because "the enormous temperatures and pressures at such depths should allow rock to dissipate stress by flowing quietly rather than fracturing suddenly..."

A closer look at the earthquake problem is addressed in relation to the location of the deep earthquakes. Since the original work on earthquakes, much data has been collected to warrant a re-examination of the plate-tectonics interpretation of subduction (Table I). Benioff zones are generally depicted as 100-km-thick slabs entering the lithosphere at a shallow angle and gradually curving to between 60° and 75°. The progenitor of Benioff zones proved that there were two separate segments. The upper portion of the Benioff Zone within the lithosphere is being compressed and dips between 20° and 45°. The Benioff Zone in the upper mantle, or strictosphere, is in tension and dips between 50° and 75°. The two are mutually exclusive in that they do not form a continuous slab and might be offset as much as 100-200 km by the asthenosphere.

Seismic coupling is used as a quantitative measurement of the interaction of plates at the subduction zone. The size of the earthquakes shows that the largest earthquakes occur in young lithosphere at fast convergence rates. In zones of old lithosphere and slow convergence rates, there are almost no earthquakes. Also, plates are largely uncoupled below 340-km depth, where the first phase change is from basalt to eclogite. Cross sections of earthquakes show the two different dip angles and segmentation for the New Hebrides, the Mariana, the Tonga-Kermadec, and the Kuril trenches. The subducting plate has been known to be decoupled at depth since the phenomenon was first noted.

Earthquake epicenter and hypocenter diagrams for the world's active margins are in Appendix B. The reader should be aware that the size of the dots on the diagrams is greatly larger than what one would see on a smaller scale diagram of the same region. For the diagram of the Mediterranean and Indian region, a breakout of the Aegean Sea is presented below that one as a demonstration. The Aegean Sea diagram also delineates a circular structure.

B. Seismic Stratigraphy

(with Dong Choi)

Seismic profiles show the locations of the reflecting rock layers, or horizons. The thickness of that rock layer is expressed as a function of the two-way travel time for the seismic wave to pass through it. Because typical travel times have been recorded for the different types of rocks, this is a tool to predict what type of rock layer is being recorded without having actual rock samples in-hand.

In the data analysis, the sequence stratigraphy established by the Exxon group was adopted. Practically, the seismic data interpretation was performed by: 1) carefully tracing geologically meaningful reflectors, 2) extraction and characterization of unconformity-bounded stratigraphic units, and 3) tying the interpreted stratigraphic units and profiles with all available geological and geophysical information including drilling, structural, stratigraphic, sedimentological, gravitational, magnetic, satellite altimetry, and nearby onshore geologic data. In this process no attempt was made to accommodate the interpretation to any pre-existing structural models.

Because the seismic profiling record the crustal section by continuously imaging the physical property contrast of the rocks which does not necessary correspond to geological features, the seismic interpretation requires a highly integrated knowledge of geology and geophysics. Wide experience in field geology is essential together with the in-depth understanding of acoustics, as well as experience in the application of seismic interpretation to practical works, such as mineral and hydrocarbon explorations. Of particular importance is that, unless supported by controlling data (such as core samples, dredge hauls, or a grid of profiles), the interpretation must be left open for further improvement or revisions. In most cases, the seismic lines across the trenches are isolated and not taken in a grid. Furthermore, very few drilling data are available from the deep sea to confirm the interpreted profiles and the real composition of "oceanic crust" (no deep sea drill holes ever penetrated basalt, Layer II, to reach Layer III, the real "oceanic crust"). A transect across a trench has never been drilled down to basalt in the entire history of the DSDP and the ODP. Leg 60 at 18°N latitude attempted, but was not successful. Leg 125 also failed at 31°N latitude. This innate incompleteness of the available deep sea seismic data and indirect form of information of the crustal section have produced a plethora of speculative interpretations, many of them being geologically inconsistent to accommodate their models. As seen in our interpretations, however, it is clear that geology under the ocean can be interpreted based on the laws of geology firmly established on land; land and ocean are governed by the same geological rules.

Exaggeration is avoided and natural scale (vertical to horizontal scale, 1 to 1) is used wherever possible, because the excessive vertical exaggeration tends to produce distorted image of the real geology under the oceans.

Many multi-channel seismic traces have been interpreted since they were first collected. Using more updated methods, they have been re-interpreted by Dong Choi. Thanks are due him for all of those presented herein. The interpretations are in Appendix C.

C. Satellite Altimetry

Many regions remain unsurveyed, especially in the southern hemisphere. Because that area is at least as large as that of the northern hemisphere ocean basins, something had to be done to include that area in any kind of meaningful discussion about structure of the ocean floors, a discussion which will hopefully lead to a more robust explanation of Earth's tectonic events leading to the current geomorphology.

Satellites were determined to be the most expeditious means of collecting data in unsurveyed regions. The Seasat was used for a brief time, giving incomplete Earth coverage, that completeness being a function of the provider. That effort led to various satellite altimeter studies. A comparison of the Seasat with first, the DBDB-5 data for the North Atlantic Ocean basin showed that the satellite altimetry data and the bathymetry agreed almost exactly, with the altimetry data continuing fracture zone trends even though the trends are overprinted by such features as the Sohm Plain.

The GEOSAT altimeter was launched to measure the altitude above the sealevel very accurately, and the sealevel is related to undulations in the gravity geoid. The GEOSAT was allowed to complete its mission in the 1980s. A high-pass filter was applied to the geoid data at NAVOCEANO which produced only the highs and lows in the sea surface during the early 1990s. A later comparison of that same SASS data set (Figure 10) to the GEOSAT trends (Figure 11) also showed a remarkable similarity, so much so, that the determination was made that the stand-alone GEOSAT structural trend data set was enough in the regions of sparse bathymetric coverage to be useful in studying regional tectonics.

While bathymetry cannot be reduced from the gravity at this time, the gravity data show the intangible, invisible trends that are reflected in the bathymetry, which shows the tangible, visible trends. This means that the gravity trends are related to the structural trends, and that the GEOSAT results can be used for a basin-wide tectonic interpretation.

The charts included herein were produced from the high-pass filtered GEOSAT charts at NAVOCEANO to show basin-wide structural trends for the Atlantic, Indian, and Pacific Ocean basins (Figures 11, 13, and 16).

The following sections are the ocean floor features. They are listed by primary and secondary geomophologic forces. A history of their discovery, general tectonic thoughts about them are presented, and a summary at the end of each chapter about what we have discovered from the other discoveries is presented. Following the presentation of the different feature types, a more in-depth discussion combining all the different facets is presented. From that you may draw your own conclusions as to the tectonics necessary to produce the present world.

D. Geodynamics

In the study of geodynamics, geophysicists have produced many mathematical models to predict what is happening beneath Earth's surface. Essentially thought to be driven by heat flow, this is an integral part of the plate-tectonic hypothesis in that huge, heat-driven conveyor belts carry the ocean floor from the spreading ridges to the convergent margins. The heat is thought to be remnant heat from Earth's formation and the heat provided by the radioactive decay of unstable isotopes in the interior.

Earthquakes, volcanoes, and the plate movement are all surface signs of this heat. Most are produced by magma chambers over the asthenosphere in the form of thermal plumes and hotspots. In the early 1980s, Woodhouse and Dziewonski showed this idea to be fallacious when they delineated major zones of heat worldwide. This idea was expanded upon in the late 1980s to a more refined model which gave the first indication that hotspots per se did not exist, rather they were hot lines. Since that time D.L. Anderson and Hetu Sheth, in 2003, have totally refuted the idea of thermal plumes and hot spots.

We will discuss these possibilities within.

OCEAN FLOOR FEATURES
MIDOCEAN RIDGES

The midocean ridges have been called the largest features on the planet, and they are active. They begin the introduction to totally surveyed ocean floor features:

The first, and most prominent, feature associated with Earth-moving tectonic events was discovered very early in the history of ocean floor study. The midocean ridge (MOR) was discovered in 1855 when Matthew Fontaine Maury drew the first bathymetric chart of the North Atlantic and called it a "great shoaling" middle ground. Subsequent cruises, such as cable-route surveys by the HMS GORGON and the HMS CYCLOPS, added to the data base. The HMS CHALLENGER Expedition of 1873-1876 found and traced the MOR.

The Navy Hydrographic Office (HYDRO; former name of NAVOCEANO) compiled a bathymetric update from 50°N to 40°S latitude in the late 1800s. In 1956 Maurice Ewing and Bruce Heezen found that small earthquakes, found near the ridge axis, outlined a steep-walled linear valley following the crest of the MOR in the Atlantic. Heezen and Marie Tharp had been coming to HYDRO with shopping carts. They would roll out the fathograms across the entire North Atlantic basin. From these Heezen constructed six profiles (Figure 4), which showed the great "North Atlantic Shoaling Ground," which would become the Mid-Atlantic Ridge. Edward Bullard discovered that the material at a shallow depth beneath the ridge was hotter than the surrounding area (also in 1956). This information was assembled and published as the world's MOR system and was presumably used partially in the construction of Figure 1.

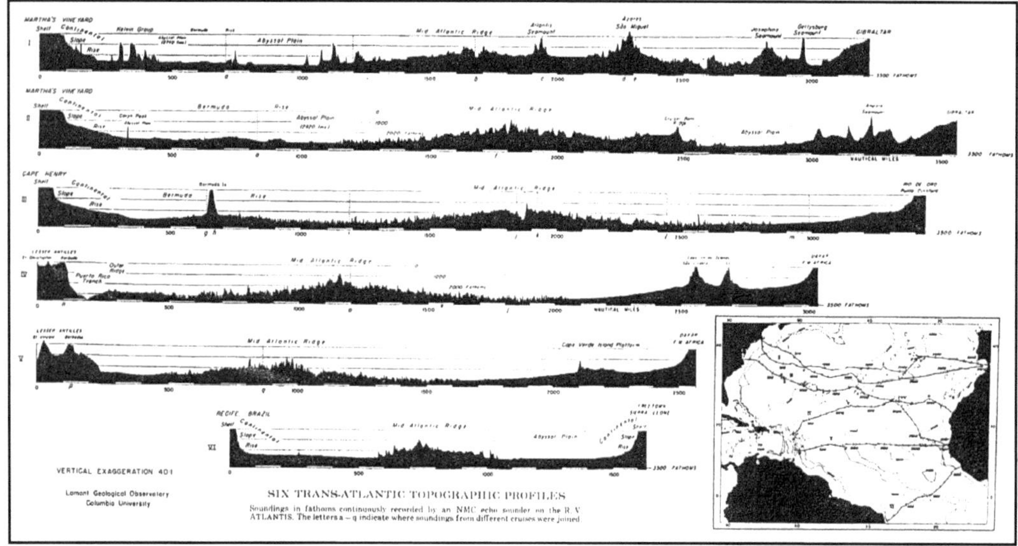

Figure 4. 6 profiles across the Atlantic basin compiled by Heezen, Tharp, and Ewing in 1959.

Bob Dietz and Harry Hess found the rift zone to be geologically active, the stress state theorized to be everywhere in tension along the MOR because of seafloor spreading. The MORs provide the "ridge push" part of the plate tectonic equation. The MOR is not spreading everywhere at once, only where active magma resides. All rocks there are extremely young and basaltic in nature from a geologic standpoint. The MOR was found to be in tension where the focal mechanisms of earthquakes show normal faulting to build an axial valley with vertical walls. Viscous material, rising in this vertical walled cleft, adds magma to the spreading ocean floor. This is where new oceanic crust is produced, so the basalt is fresh. As the crust moves away from the MOR it cools and subsides so that cracks form and the features become blocky and jumbled. This imprint gives short tectonic spreading fabric (TSF) fractures parallel to the ridge axis. The new ocean floor forming there records the magnetic field at the time of its formation. This is the spot originally thought to be the site of seamount formation, the thought being that thousands of seamounts would be found on the ocean floor, most of which had formed here initially and drifted to their present locations.

The MORs are segmented, being offset by transform faults. Through various tectonic processes, the segments may become offset so that they overlap. The overlapping spreading centers (OSCs) are non-rigid discontinuities. The axial depth deepens by several hundred meters, the geochemistry of the erupted lava changes, and the fine scale structure of the ridge changes. OSCs vary in segmentation length to produce second-, third-, and fourth-order discontinuities. The shorter length, 10-50 km, fourth-order OSC occurs as very small axial summit graben offsets that are barely detectable bathymetrically. TSF produced in this tensional environment fit the description of OSCs.

Segmentation of the MORs is responsible for the pinch-and-swell geometry whose segments average 50-300 km long. The segments are sequentially scaled, beginning with the OSCs. The next larger segment, separated by less prominent ridge-transverse depressions, have been called transfer zones, devals (narrow axial summit depression), and accommodation zones. The largest have been called transform faults, or ridge-transverse fault zones. These segments actually range between 15 and 500 km in length, and they are a minor player in the overall scheme rather than what they originally referred to by Tuzo Wilson (1965).

On-ridge seamounts are formed at the crest of the MOR and sometimes continue to grow as they subside. In the Atlantic on-ridge seamounts exist primarily from the Hayes Fracture Zone to the north. On the basis of morphology and structural evidence, most small seamounts probably originated near ridge-transform intersections and fracture zones and are round, although the shape is thought to be determined by hydraulic resistance to the flow of magma through the edifices. Their life-span is very ephemeral for the most part. On-ridge seamount formation is a function of the magma chamber and supply beneath the rift system. A slow spreading system such as the North Atlantic produces very few seamounts. A medium spreading rate such as the Juan de Fuca Ridge will produce en echelon volcanic chains. A fast-moving system such as the East Pacific Rise produces many more seamounts. Seamount volcanism was more active in the Eocene. Beyond the Eocene the number of seamounts decreases, but the volume increases. The term MOR is not true for every ocean because the ridge is sometimes off to one side of the basin, such as the East Pacific Rise.

The northern Mid-Atlantic Ridge, hereafter called simply the MAR, stacked profiles show the MAR to be shoaler at the crest tapering to low abyssal plains on the flanks. The various highs along the profiles are blocks, ridge-formed seamounts, or sundered plateaus. This is the standard acceptance of the ocean basin feature. Superimposed profiles of the MAR at 17°30'N latitude and the front range of the Rocky Mountains at 39°N latitude show such remarkable similarity, that they must have been produced by the same tectonics.

In a cooperative effort, researchers have surveyed a large portion of the northern MAR. Mohns Ridge (Figures 3-17, 5, and 6) in the Greenland and Norwegian Seas extends from 5°W to 7°E longitude (scan forward to Figure 59 to see how this is diagramed). This segment, spanning from 0° to 3°E longitude, has been covered near-totally by SeaBeam surveys.

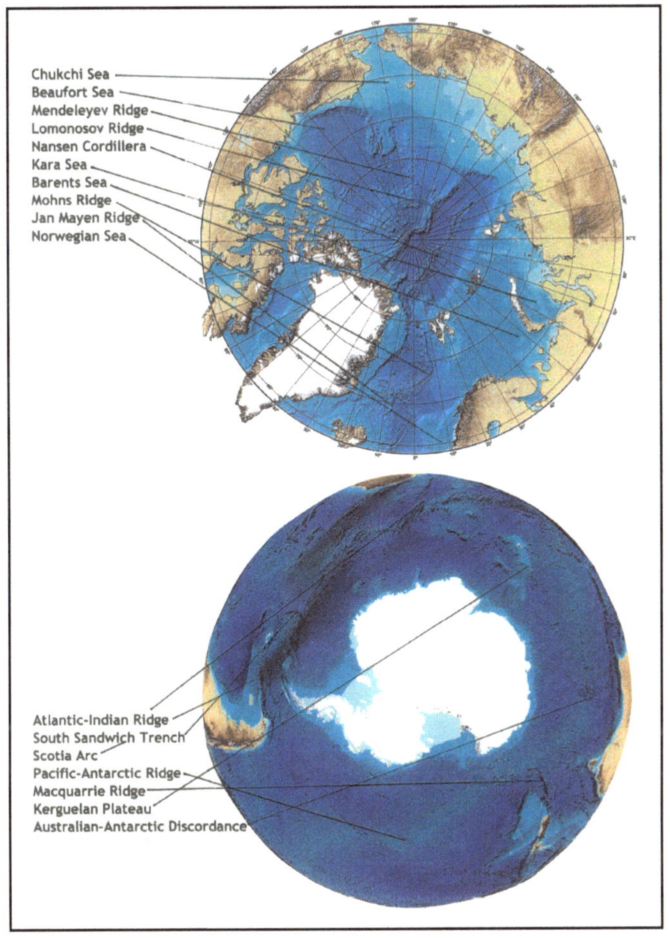

Figure 5. International Bathymetric Chart of the Arctic Ocean (2001). The effort is still in progress for the Antarctic Ocean. GEBCO has a 1.0-minute grid and that is on the bottom of the figure. Obviously, not enough data exists in the world to support a 1.0-minute grid of this region, but we shall take it at face value.

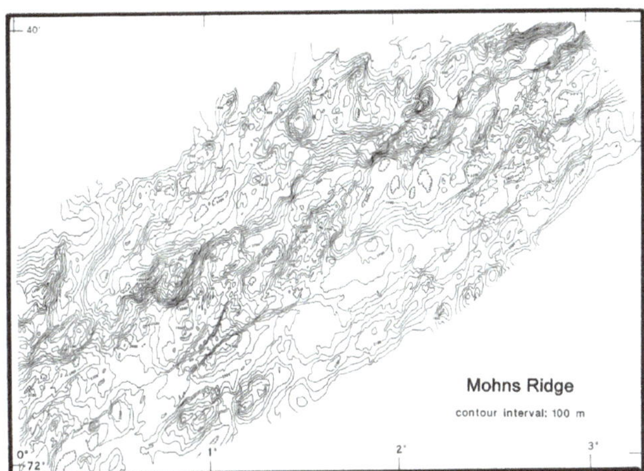

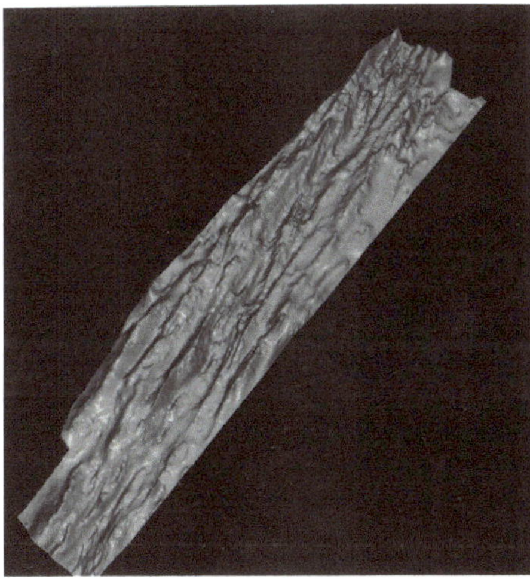

Figure 6. Mohns Ridge in the Arctic Ocean on the Mid-Atlantic Ridge, surveyed with the SeaBeam system and contoured at a 100-m interval. Notice in the 3Ds that the FFF/TSF is not parallel to the ridge axis, instead being at an angle. At that, they are parallel to each other.

From SW--NE, the axial valley is bounded on the flanks by coalesced seamounts and ridges that range from 1500--2000-m deep. The 3200--3400-m axial valley is disjointed, segmented, and offset 30° from the overall strike of the MOR. The en echelon valleys are bounded on the ends by large seamounts. The valley floor has many volcanoes and ridges. The gross morphology is nothing like that predicted by plate tectonics.

Iceland lies to the south. Iceland has at least four rift zones, so we may say that, in keeping with the spirit of plate tectonics, it has a diffuse spreading center. Magma wells up in parallel to sub-parallel fissures, faults, and fractures for the entire length, a region from latitude 55°N to 70°N of the MAR. As we have seen, two of the volcanoes in this spreading center produce up to 75 km^3/yr of lava. The continuous MOR of the MAR at Iceland not only diffuses but it also disappears for over 180-km north of the island to Kolbeinsey Ridge and 330 km NE to Jan Mayen Ridge. This may or may not be a function of the deep volcanic pile created by Iceland itself. The portion of the MAR south of Iceland is the Reykjanes Ridge.

This segment of the MAR that has been used constantly to prove the plate-tectonic theory is adjacent to Iceland. The Iceland Plateau had been thought to have been formed as part of the British-Arctic plateau-basalt province, which had broken up during the initial rifting of the Atlantic Ocean and sunk below sealevel. Because of Iceland's position on the MAR, investigators decided to have Iceland formed on the ridge-crest by volcanic eruptions. In a later explanation, a hotspot was felt to be the creator. Yet later work theorized at least three rifts through Iceland and no extensional tectonics (read seafloor spreading) because the neo-volcanic zone (new rocks) overlay Pliocene stratigraphy (older rocks). The debate papered the press, and in 1992 the magnetic lineations used to "verify" seafloor spreading on the Reykjanes Ridge were refuted by Bill Agocs and Karoly Kis.

As previously noted, the Iceland Plateau seems to have been continental at one time, at least 700 Ma. Spreading is possible only in rift zones, so no spreading occurred in this part of the Atlantic basin. New lavas can hardly overlay older unless the implanted rocks remain stationary; that is, no horizontal movement occurs.

The MAR between 37° and 35°N encompasses the Project FAMOUS area and the Oceanographer Fracture Zone. From the southern end the rift axis leaves the ridge-transverse fault area at exactly the nodal basin. The first 55-km segment of the MAR on a 015° azimuth is a very pronounced rift axis. Fabric is imprinted on the high 1830-m blocks, or horsts, forming. A noticeable lack of axial volcanoes exists on this segment. The second segment, offset 46-km to the east, is 65-km long and has an equally, if not more so, pronounced rift valley ranging to 3110--3475-m deeps. No volcanoes are present at this contour interval. Here again large, 1800-m relief horsts exist on each side of the rift valley. The third segment is offset 15-km east. At this contour interval, many axial volcanoes are present. While fabric is being imprinted, it is doing so at a lesser rate than further north, presumably because there is less of a magma outpouring for crustal creation. The fourth segment is offset 22 km and is 44 km long. It, too, is rife with axial volcanoes. Apparently the MAR meanders from one ridge-transverse fault to another, or there are more ridge-transverse faults than previously thought.

Photo mosaics of the FAMOUS area of the MAR show the axial valley floor strewn with pillow basalts and broken lava rock debris. This axial rift was found to be a narrow trough approximately 2-km wide, when the first multibeam sonar was collected and presented. The axial rift valley was characterized by low relief, linear ridges on the floor and narrow flat-topped terraces and benches with outward facing antithetic scarps on the walls from 36-37°N.

The portion of the MAR south between the Oceanographer (35°N) and the Hayes (33°N) fracture zones is also segmented. The first 70-km segment north of the Hayes has a clean axial valley which is 2200-m deeper than the surrounding walls (Figure 7). The second 82-km segment is offset to the east by an unnamed 70-km long ridge-transverse fault. The 2000-m deep axial valley is loaded with volcanoes. The third segment is offset 30-km to the east by yet another ridge-transverse fault before continuing to the NE for 93-km and joining with the Oceanographer Fracture Zone. No readily apparent axial volcanoes appear in its 550-m deep axial valley.

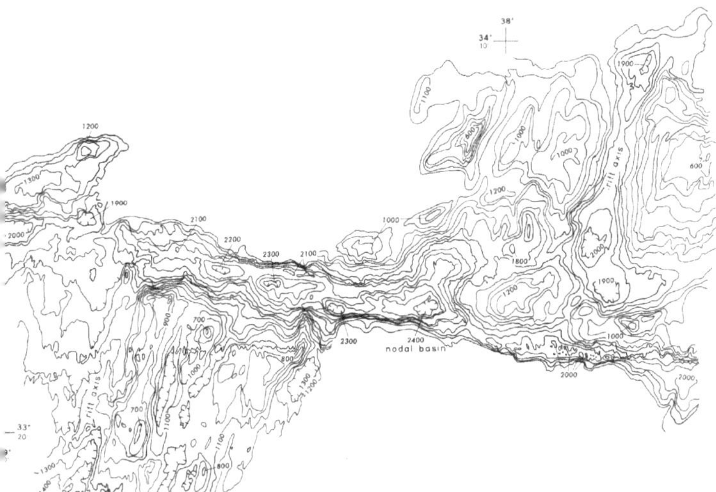

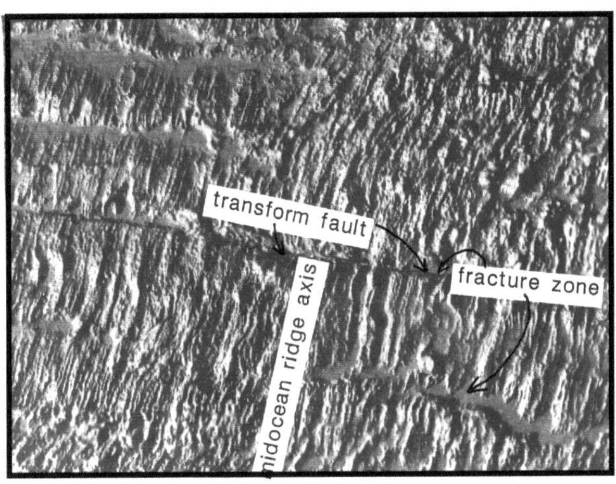

Figure 7. The Hayes Fracture Zone (top) transform region of the MAR from multibeam sonar surveys at a 100-fm contour interval. The nodal basins define the limits of the transform fault. The MAR around the Atlantis Fracture Zone based on multibeam sonar surveys (bottom). The 3D of this region shows that not all fractures pass through the MAR, leading one to think that possibly the fracture formed first and was overprinted by later deposition. This 3D shows the ridge segmentation and offsets caused by the braiding of the fractures.

The Atlantis Fracture Zone lies to the south of that at 30°N (Figure 7). This 3-D shows that some fracture/transfer axes cross the Mid-Atlantic Ridge, some braid, and some don't cross at all. Considering the fact that this diagram covers about 175,000 km^2, this is no small section of the ocean floor.

Switching ocean basins, the MOR for the Pacific Ocean basin, the East Pacific Rise, is on the eastern side of the basin (Figure 3-6). The East Pacific Rise has been extensively surveyed and described by SeaBeam sonar by such programs as RIDGE and RIDGE 2000. The results are easily located on the Internet, so they are not presented here, other than to note their existence. It shows a mixture of flat terrain to overlapping spreading centers. Some of the hummocky terrain is explained by the presence of black smoker chimneys, which may be up to 10-m high. The East Pacific Rise has no clearly defined rift zone.

From a geomorphology standpoint, the East Pacific Rise runs ashore at the southern tip of Baja California and continues north in the form of the mountain ranges to the west of the Rocky Mountains. It re-emerges to the north in the form of the Juan de Fuca Ridge, an offset by the San Andreas Fault of some 3000 km, a feature which is called a transform fault. Once again, this is the central, active portion of a fracture zone and seafloor spreading is parallel to the feature. This will discussed later.

Backarc basins have been hypothesized to be opening by spreading, or "unzipping" from south to north as the case may be. A salient example is the Mariana Trough, or backarc as it is known (Figure 3-11). A series of profiles constructed for the length of that basin, located in the region of the Philippine Sea, reveals that no spreading center exists, and that the TSF is diffuse throughout from west to east, and that all of the ridges are of more-or-less equal height. This basin has been completely surveyed with multibeam sonars, so no doubt exists in the interpretation.

The tectonic interpretation of the function of the MOR has been in a state of flux. Sidescan sonar imagery has not only given high-resolution coverage but also much continuous coverage over thousands of square kilometers. Observations based on actual data include, aside from the transform faults every 35 km or so in certain regions, the occurrence of non-transform ridge axis discontinuities such as (OCSs) and en echelon faults, fissures, and fractures that parallel the spreading axis. This phenomenon reflects Stokes' Law; essentially Newton's Second Law of Motion. Magma flow is indicated to go along-strike on the ridge instead of away from the ridge. Poiseuille flow patterns are flow-parallel shears between different velocities, such as in lava tubes, glaciers, and the MOR, and the parallel cracks, in turn, are the geomorphic expression of Poiseuille flow. The TSF has by now become the ridge-parallel fault-fissure-fracture (FFF) fabric. The Reykjanes Ridge, the FAMOUS area, the Juan de Fuca Ridge, and the East Pacific Rise laminar flow features are offset at the recognized transform faults, with several of the transform faults overprinted. The fissures, faults, and fractures continue unbroken in a N--S direction.

The MORs are everywhere characterized by the FFF systems, even out onto the flanks. These are stretching lineations, and they continue onto the continents in many cases, such as the East Pacific Rise. This particular FFF bathymetry is characteristic of all tectonic belts, so that description includes not only MORs but also aseismic ridges, linear island and seamount chains, oceanic active volcano arcs, backarc basins, and foldbelts. Essentially, the magma flows along strike, not perpendicular to it and out into the world. Much leaks out to overprint the 1.5 Ga rock already in place.

A simple exercise in morphometry gives the lengths of the MOR sections: the African Rift system (3840-km; Figure 3-1) and continued through the Red Sea (2000-km; Figure 12-20), Arabian Sea (1280-km), Carlsberg Ridge (2000-km; Figure 3-2), Mid-Indian Ocean Ridge (4200-km; Figure 3-3), Southeast Indian Ocean Ridge to the MacQuarrie junction (6500-km; Figure 3-4), Pacific-Antarctic Ridge from the MacQuarrie junction to the Eltanin Fracture Zone (3500-km; Figure 3-5), and up the East Pacific Rise (10,000-km; Figure 3-6). While in the Pacific basin the MORs for the Nazca basin (5500-km; Figure 3-7), the Galapagos Ridge (2300-km; Figure 3-8), the Gulf of Alaska (1500-km; Figure 3-9), the Lau-Havre Ridge (2700-km; Figure 3-10), and the Mariana Trough (1400-km; Figure 3-11) conclude that basin's MORs. The Indian Ocean basin MOR from the Bouvet junction to the Mid-Atlantic Ridge is first the Southwest Indian Ocean Ridge (4000-km; Figure 3-12) and the Atlantic-Indian Ridge (2000-km; Figure 3-13). A portion continues to the Scotia basin, that containing 1850-km (Figure 3-14). The Mid-Atlantic Ridge is more-or-less 18,400-km long (Figure 3-15), depending on where you stop in the Arctic Ocean. This figure includes the Reykjanes (Figure 3-16), Kolbeinsey, Mohns (Figure 3-17), and Lomonosov Ridges. Finishing the MORs recognized by most authorities, the Cayman Trough at 1100-km (Figure 3-18) gives a grand total of 77,070-km of MORs. This total ignores the recently proposed MOR dividing the India-Australia basin, which would only increase the value by several thousand kilometers.

Seafloor spreading occurs in both directions perpendicular to the ridge, so we have 154,140-km of potential spreading at any one time, depending on the magma supply. In order to prevent an expanding Earth, that much must be taken up by engulfment or shrinkage.

FRACTURE ZONES

Fracture zones, as zones of weakness in Earth's crust, are associated with mid-plate seismic events, suggesting that their provenance may be related to the buckling and fracturing of the crust caused by the release of mid-plate stress. At any point of weakness at any time magma leakage through the fracture zone can produce seamounts or islands. While the original thoughts had the seamount chains to be age sequential; that is, to progress from younger to older, that has since been proven not to be the case. Seamount chains are the result of normal tectonic activity, and one finds all-aged seamounts and islands intermingled. While this discovery precluded the idea of hotspot generated seamount chains, the data have received very little attention. The presence of a fracture zone can then be determined by the presence of linear seamount chains along with the familiar strike-slip ridge-valley configuration.

In the plate tectonic hypothesis, fracture zones, locked and in place rather than active, show the direction of seafloor spreading at the time they were imprinted, or formed.

Observation of what the bathymetry/topography shows through the various means such as multibeam bathymetry and satellite altimetry can help fine-tune the original definition. Fracture zones were defined as extensive linear features of unusually irregular topography of ocean floor characterized by more than one kind of feature such as seamounts, asymmetrical ridges, troughs, or escarpments having age and regional depth contrasts across the younger end of fracture.

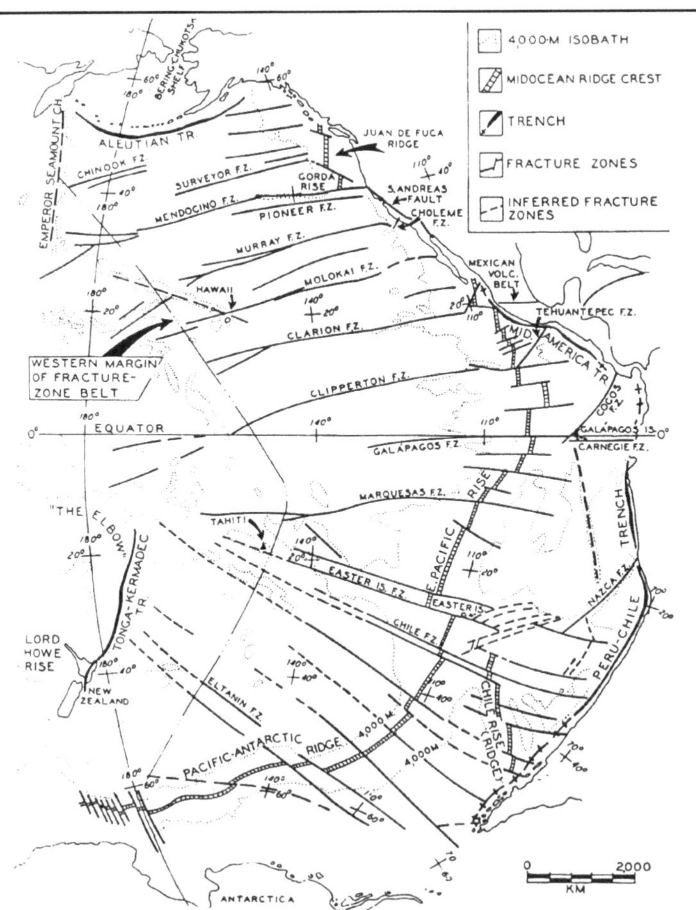

Figure 8. 1972 diagram based on then-available bathymetry of the Pacific basin. The Meyerhoffs realized at that time that the fracture zone azimuths could not possibly show the direction of seafloor spreading, instead showing a fanning pattern from the western basin. This was completely ignored.

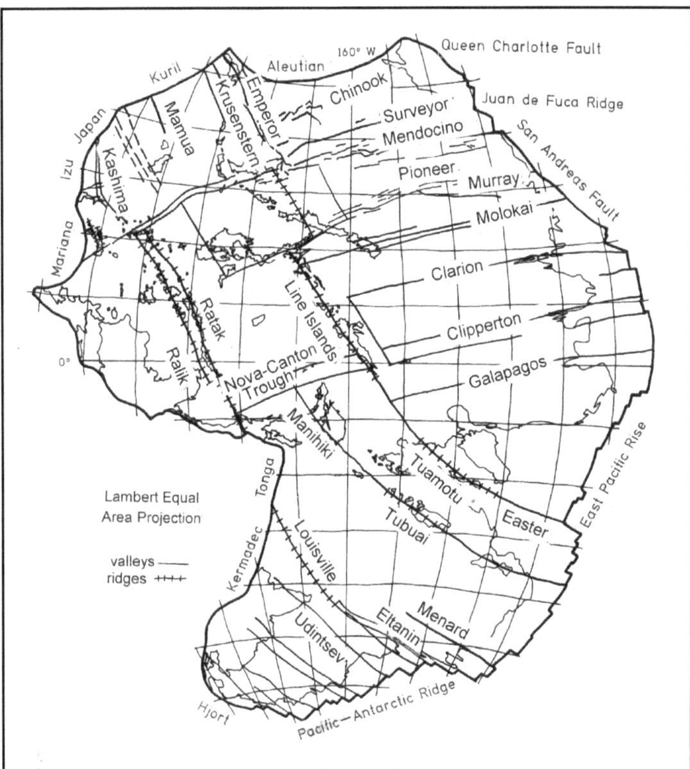

Figure 9. By 1988 enough good bathymetry was in place to draw this diagram of the major trends in the Pacific basin. By ignoring the 43-Ma "bend" while realizing that many of the seamount chains were extensions or parts of the fracture zones, and continuing the lineaments, it was possible to see that the lineaments intersected orthogonally. Still the data were ignored, and it became increasingly difficult to get that information published in mainstream journals.

This definition had nothing to do with the tectonics except that, through common usage, they were felt to be either active or fossil plate boundaries. Within an active plate boundary strike slip faults predominate. The topography was thought to be the result of differential vertical adjustment with possible extension and compression components. Theory has the direction of seafloor spreading shown by the fracture zones, a concept which allows one to infer an ocean basin's history. If the lithosphere is in motion, the direction is the same on both sides of the fracture.

By the time the World Chart Group had produced the first all encomapassing ocean floor maps in the early 1970s, the observation that, for the Pacific basin, the fracture zones could not possibly show the direction of seafloor spreading (Figure 8). This was noted by the scientific advisory group from several of the universities, a finger-pointing session ensued, and the matter was promptly dropped (Frank Marchant, World Chart Group Head, pers. comm., 1989). The fracture zones defined a fanning pattern across the basin, all converging in the west-central basin. been published OSP data not only verified the original fanning pattern, it introduced the concept of intersecting fracture zones.

For many years now the fracture zones have been known to braid, meander, and splay. Using a highly refined bathymetric surveying system, fracture zones are known to exist that are orthogonal to the fracture zones used in the early hypothesis (Figure 9). The San Andreas Fault is a perfect example. The Mendocino and Murray fracture zones have been proposed to abut the San Andreas Fault at a 90° angle. This fact alone should have raised a red flag. Later research has the Mendocino and Murray fracture zones coming ashore, the Mendocino going all the way to Yellowstone National Park in Wyoming. In fact, most of the fractures do not stop at the continental margins; they continue on through the continents to emerge in the opposite ocean basin. Some circle the globe. Disregarding the proposed elbows at 43 Ma led to a checkerboard pattern for the entire Pacific basin. And, it showed the linear seamount chains to be extensions of the fractures. This was of course a physical impossibility within the constraints of the plate tectonic hypothesis.

1. Atlantic Ocean Basin

But that diagram was based on data that was not to be the final bathymetry. As an example, by the mid-1980s NAVOCEANO had completed the surveying and charting for the northern Atlantic Ocean basin. The charts were photographically reduced and pasted on a grid, producing a "superchart." From this it was possible to compile and publish a fracture zone locator diagram (Figure 10) based on near-total bottom coverage. Fracture zone trends in the Atlantic Ocean basin had been thought to lie mostly WNW--ESE in

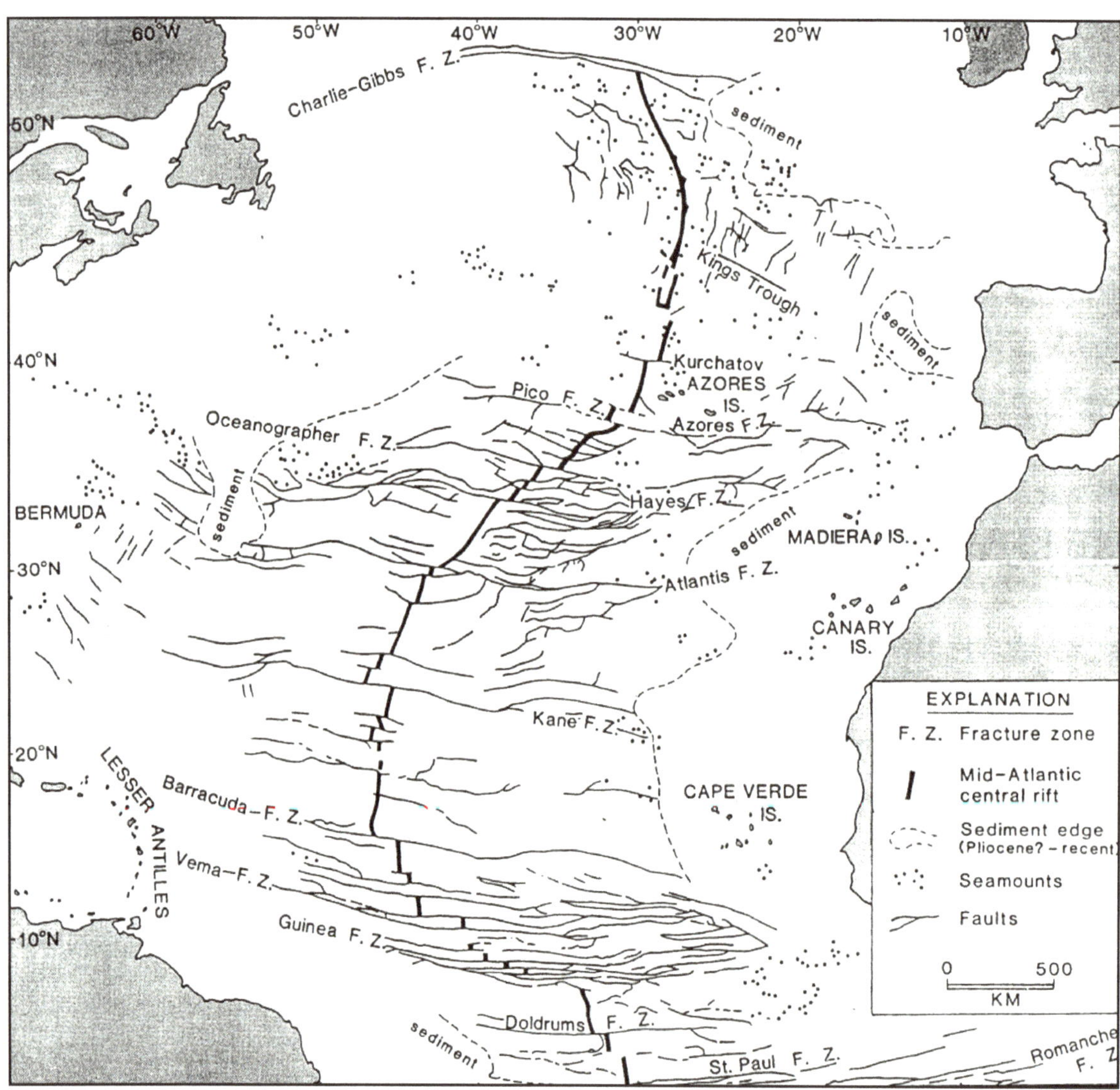

Figure 10. 1989 multibeam sonar-based tectonic diagram of the North Atlantic ocean basin on a Mercator projection showing all of the fracture valleys and the seamount locations. Note that the region between the Pico and Charlie-Gibbs fracture zones, while rife with fabric, has no crossing fractures and holds most of that basin's seamounts. While the number of seamounts may seem small compared to those given by others, the reader is reminded of the definition of a seamount as opposed to a hill or a knoll. I have even seen the chimneys/black smokers listed as seamounts.

parallel-to-sub-parallel straight lines. The fracture zones did not lie at discrete intervals. They were not parallel-to-sub-parallel. Fracture zones are not straight. They splay, merge, and stop and start indiscriminately within the constraints imposed by the lithosphere and stress regime. Fracture zones became "fracture zone swarms."

Fracture zones were defined as extensions of the transform faults on the midocean ridge spreading centers. This idea is not even remotely accurate; at least half of the fracture zones in the Atlantic basin are not even associated with transforms. They do not cross the midocean ridge. Instead, they are overprinted by it. This means that the fracture formed before the midocean ridge, because the fractures continue on both sides of the ridge. The fracture zones in the Pacific basin that approach North America are not associated with any ridges, yet the magnetic stripes show them to be still forming from the east.

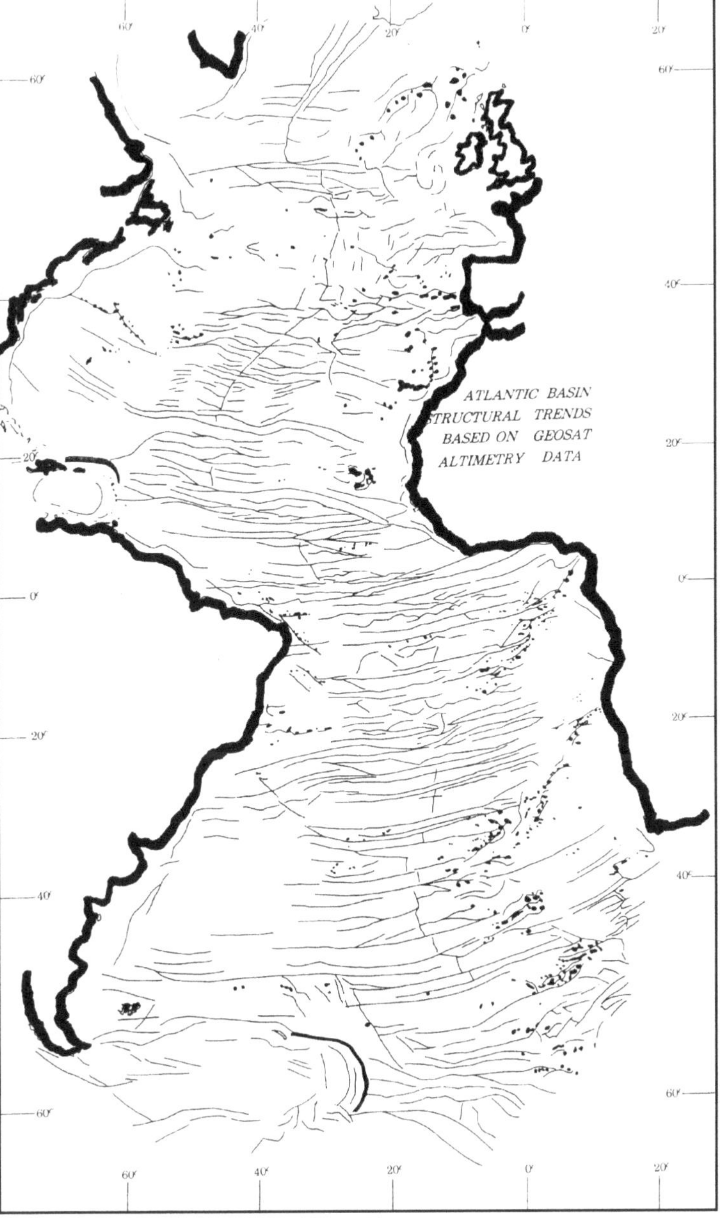

Figure 11. 1995 GEOSAT high-pass filtered diagram of the Atlantic Ocean basin on a Mercator projection. Comparing this figure to Figure 10 reveals a remarkable resemblance, so much so that one could conclude that the high-pass filtered satellite data is the equivalent of structural trends. This opens all of the ocean basins for tectonic scrutiny rather than relying on snippets of information

A perfect example lies on the MAR between Iceland and the Oceanographer Fracture Zone. Two fracture zones have been proposed for this region. By definition they must join transform faults and cross the MAR. Neither the Faraday nor the Maxwell "fracture zones" cross the MAR (Figure 10). They are, in fact, overprinted by it. Similarly, the same phenomenon occurs further south on the MAR at the Atlantis Fracture Zone (Figure 7) where lineations appear on both sides of the MAR while not actually passing through in the bathymetry. This is not a rare occurrence.

A study comparing the Figure 10 to the GEOSAT diagram (Figure 11) for the North Atlantic basin produced 52 fracture zones with 13 crossing the MAR. The alignment of Atlantic basin fractures has always been generally thought to be EW. Thirteen transform faults were found to cross the Mid-Atlantic Ridge between 0 and 55°N latitude. The fracture zones are absent from 40°N to 52°N between the Hayes and Charlie-Gibbs fracture zones. An additional 25 to 38 fracture zones lie off-ridge. Hence, fracture zones do not necessarily form as extensions of transform faults, apparently being controlled more by the regional stress field. Also, the idea that they may have been formed before the implantation of the midocean ridges began to take shape.

In the synopsis of the available data of the Atlantic Ocean basin, most of the megatrends are aligned WNW--ESE in the North Atlantic. Starting on the north at the Charlie-Gibbs Fracture Zone and continuing south to the Azores/Pico Fracture Zone, almost no E-W fractures exist. Instead, this region is characterized by a high number of seamounts; nearly 200, according to the superchart, in all with no particular alignment. Nowhere else in the Atlantic Ocean basin does this phenomenon occur. Nor, do they reach the

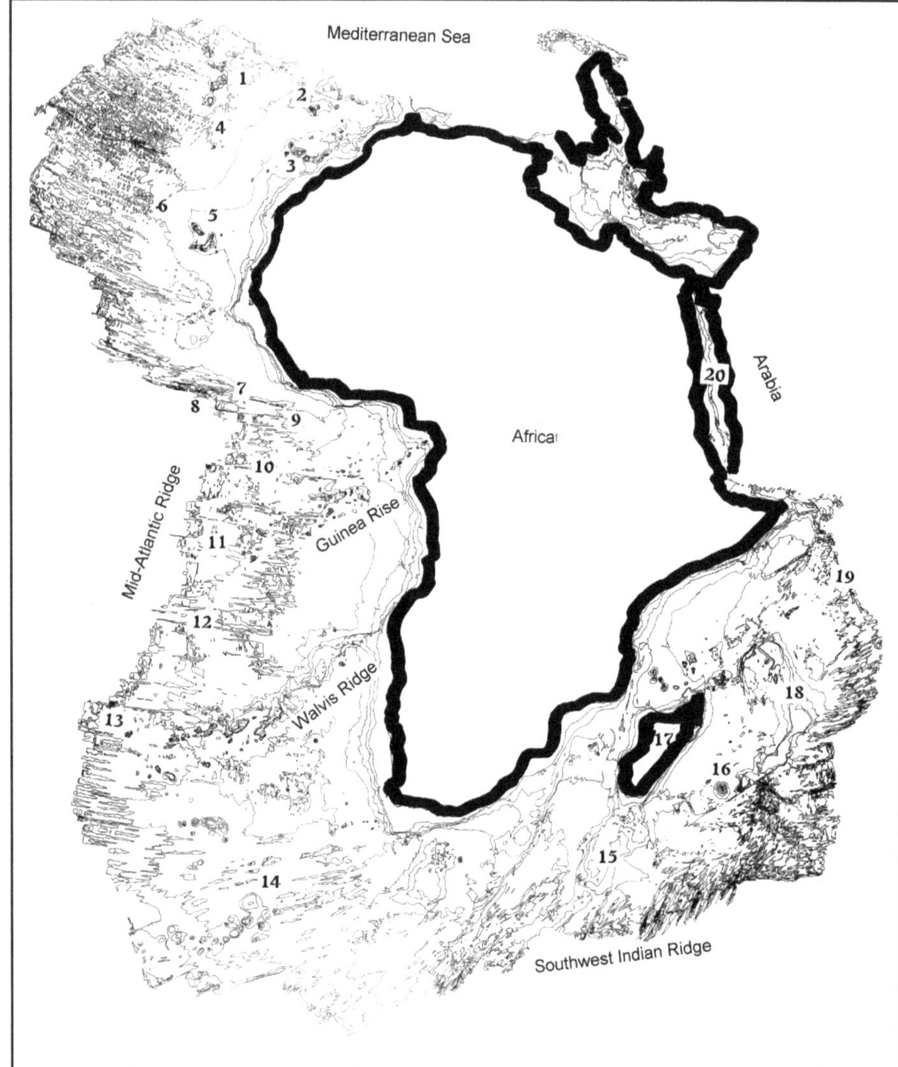

Figure 12. Based on the GSA Map MCH-069 (1989) and the DBDB-5, this diagram of the proposed African plate at a 1000-m contour interval on a Mercator projection shows: 1=Cruiser/Hyeres seamount group, 2=Madeira Island, 3=Canary Islands, 4=Atlantis Fracture Zone, 5=Cape Verde Islands, 6=Kane Fracture Zone, 7=Romanche Fracture Zone, 8=Ascension Island, 9=Ascension Fracture Zone, 10=Bode Verde Fracture Zone, 11=Bagration Fracture Zone, 12=Rio de Janiero Fracture Zone, 13=Tristan da Cunha, 14=Cape Basin, 15=Madagascar Ridge, 16=Reunion, 17=Madagascar, 18=Mascarene Plateau, 19= Carlsberg Ridge, 20=Red Sea. The reader will notice that this "plate" is almost totally surrounded by "spreading centers."

midocean ridge crest. The possibility remains that, although seamounts do show an underlying fracture in most instances, this region has a different stress field because the ocean basin is so narrow between the continental cratons.

From that point to the equator, three of the major fracture swarm/megatrends traverse the entire basin. The Oceanographer/Hayes/Atlantis swarm includes the seamounts at the distal ends near the continents. The second is the megatrend comprised of the Barracuda/Vema/Guinea fracture swarm and includes the Bathymetrists Seamounts. This feature offsets the midocean ridge for the distance of 1100 km and appears to go ashore in western Africa on the east. The western terminus on the Atlantic Ocean floor appears to be the Lesser Antilles island arc. The third is the St. Paul/Romanche fracture swarm, which straddles the equator to 3°S latitude (Figure 12-7). It is here that the megatrends change direction from WNW--ESE to WSW--ENE, and this trend becomes more pronounced the further south one looks. Fracture trends in the South Atlantic basin are generally evenly spaced and do not display the contortions provided by the counterparts in the north.

G.B. Udintsev, head of the Laboratory of Geomorphology and Tectonics of the Ocean Floor and Corresponding Member of the Russian Academy of Sciences, wrote in his 1996 report to UNESCO: "...concluded that the numerous faults which intersect the equatorial part of the MAR are varying ages and relate differently to to the structure of the rift zone in time and space...structure is anomalous and differs substantially from the adjacent north and south parts of the ridge. It was also observed that spreading in the area is clearly evident only in the narrow axial zone of the segment." Most of the results of this comprehensive survey showed that a "continental bridge" existed between Africa and South America since Aptian time (125-112 Ma).

Nevertheless, the WNW-ESE trending north Atlantic fractures exist, and we see the WSW-ENE trending south Atlantic fractures. Interestingly, this fact remains largely uncontested. However, this alone belies the concept of unidirectional Atlantic basin spreading.

On the structural diagram verified by the bathymetry, trends paralleling the midocean ridges are also noticeable. One starts on the Bermuda Rise, passes NE through the southern Sohm Plain, skirts the eastern portion of the Corner Seamounts while crossing the Oceanographer/ Hayes megatrend, is overprinted by the Mid-Atlantic Ridge, goes into the King's Trough region, and continues into Ireland, across the Irish Sea, and on through Scotland. The study does not include any further north, merely noting the existence of this trend, called the "Caledonian" in Scotland. Presumably this feature formed on land during the Caledonide Orogeny, a period which lasted from the Ordovician (about 505-438 Ma) to the Devonian (began about 409 Ma). The Bermuda Rise is known to be continental in origin with DSDP ages of 671 (Proterozoic) and 381 (Devonian) Ma. As noted, this trend also passes through the King's Trough. Ordovician (440-530 Ma) graptolites and trilobites have been found there (to be discussed in the Tectonics chapter of the Atlantic Ocean basin). The King's Trough is on the ocean floor, so we may take the liberty of assigning this age to the entire feature, and to herein name it the "Caledonian Megatrend."

Similar SW-NE trends appear in the south Atlantic, one encompassing the northern cluster of seamounts on the Guinea Rise (Figure 12), and one underlying the Walvis Ridge (Figure 12) to the south. Oil exploration reports on-line show four stratigraphic units for the Walvis Ridge and it's onshore counter-part, Namibia. The pre-rift Karoo section contains preserved Permo-Carboniferous-Triassic age samples, and this section continues offshore. The Walvis Ridge is a later addition to this base.

Last, several NW--SE trends appear in the GEOSAT structural map. The swarm on the western terminus of the Kane Fracture Zone continues southesterly and is co-linear with the eastern trends between the Barracuda and Vema fractures. Another appears at the coast of Brazil, about 25°S latitude, 40°W longitude and continues trans-basinal to 50°S latitude, 20°E longitude.

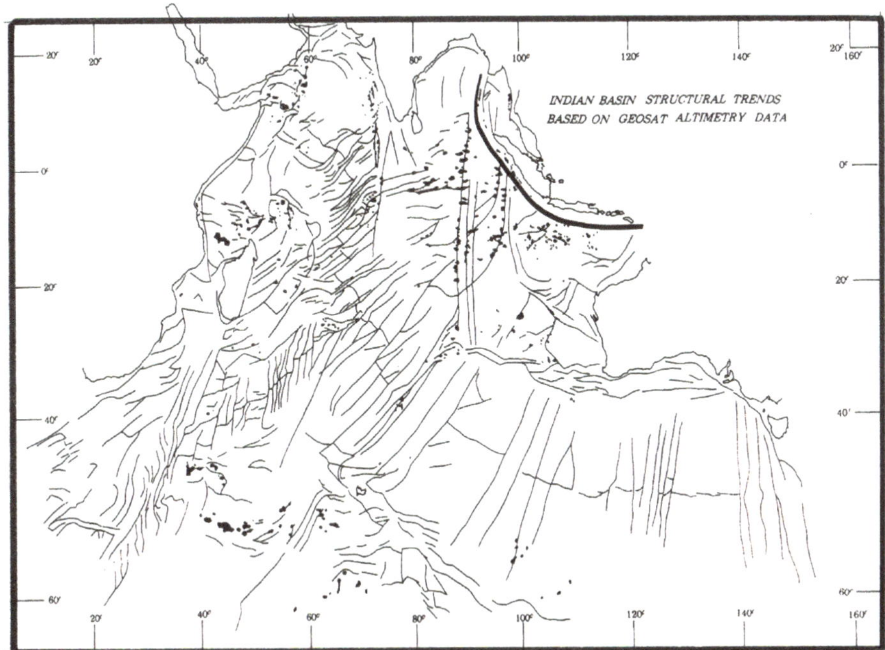

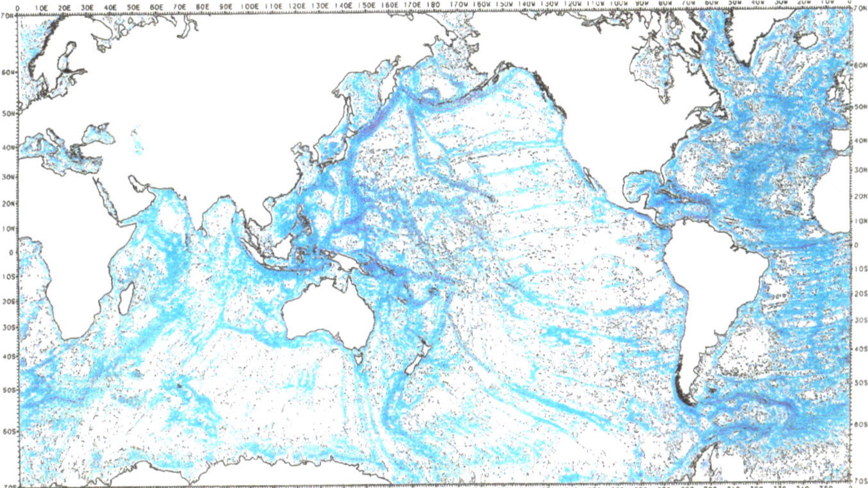

Figure 14. 1995 high-pass filtered GEOSAT structural diagram for the Indian Ocean basin to be used in conjunction with Figure 13. The bottom is a world chart of the filtered GEOSAT sea-surface height data where blue contours are positive and green contours are negative. A larger version of this chart was used in the compilation of Figures 11, 13, and 15.

2. Indian Ocean Basin

The bathymetry (Figure 13) and the GEOSAT (Figure 14) data were used to construct a basic outline of events in the Indian Ocean basin. The

Figure 13. 1995 Indian and Australian ocean basins bathymetry at a 1000-m contour interval from the DBDB-5 data base with updated contouring driven by the GEOSAT lineations to be used as a locator diagram for the text where: 1=Carlsberg Ridge, 2=Owen Fracture Zone, 3=Central Indian Ridge, 4=Southeast Indian Ridge, 5=Diamantina Fracture Zone, 6=90-east Ridge, 7=Investigator Fracture Zone, 8=Indoysian Belt, 9=Broken Ridge, 10=Arabian Basin, 11=Chagos-Laccadive Ridge, 12=Andaman-Nicobar Trench, 13=Bay of Bengal, 14=Australian-Antarctic Discordance, 15=Java Trench, 16=Lord Howe Rise, 17=New Hebrides Trench, 18=Fiji Plateau, 19=Tonga Trench, 20=Lau-Harve Basin, 21=Kermadec Trench, 22=Norfolk Rise, 23=New Zealand, 24=Macquarrie Ridge, 25=South Tasman Rise.

19
18
17
20
21
22
Australia
16
23
24
25
14

most glaring observation to be made for the Indian Ocean basin is the differences in strike and pattern of the fracture zones, especially between the SW and NW/SE quadrants. The NW fractures show a left-lateral drag effect in the bathymetry, and this includes the Carlsberg Ridge in the NW quadrant. From the triple junction near 25°S latitude, 70°E longitude to the Owen Fracture Zone at 10°N latitude, 55°E longitude, the fractures trend SW-NE. For the Southwest Indian Ridge, they trend N-S. The fracture spacing is also different. It is closer on the Southwest Indian, Mid-Indian, and Carlsberg Ridges and more widely spaced on the Southeast Indian Ridge. North and south of the Southeast Indian Ridge fractures do not exist at this contour interval. Given a little assistance by the GEOSAT diagram, the fractures appear to be similar in length-to-longer than their basinal counterparts, and they appear to stop at the Diamantina Fracture Zone, a WNW-ESE oriented feature.

Observation allows this interpretation. Fractures on the SW branch of the Mid-Indian Ridge appear to have been dragged into the morphology now existing because they point more easterly than do the fractures to the north on the same ridge.

Outside the midocean ridge segment fractures, the large fractures seem to be in the eastern basin. Several large, N-S trending ridge/trough structures lie in that portion of the basin; the 90-east Ridge, a seamount chain, and the Investigator Fracture Zone. The SW-NE trending fractures do not appear to transect the 90-east Ridge, instead they seem to have been previously imprinted and are now considered to be inactive. That same fracture deflects to the north at the Investigator Fracture Zone. This megatrend, by the way, is now being touted as the next plate boundary by the adherents of the plate-tectonic hypothesis. That is based on the enormity of the mid-basin earthquakes.

The Indoysian Foldbelt and Broken Ridge are the sites of the E-W trending fractures. The fractures between the Diamantina Fracture Zone and the Broken Ridge do not cross either of them. On a smaller scale, E-W-trending fractures appear around the Kerguelan Seamount Chain, and they appear to be co-linear with Diamantina.

The fractures to the south of Africa are aligned SW-NE and would appear to be extensions of those to the east of Madagascar were that island not in the way. The fracture pattern in the Arabian Sea gives the impression of a whorl, where they intersect the Chagos-Laccadive Ridge on the east, bend to the SW to intersect the Carlsberg Ridge, and trend generally NE in the Gulf of Oman. Another whorl appears to the south of the Andaman-Nicobar Trench off NW Australia. The whorls do not coincide with gravity geoid highs or lows.

Seamount provinces appear, once again at this contour interval, to be mostly off-ridge. The primary sites are at the Owen Fracture Zone, the Chagos-Laccadive Ridge, north and northeast of Madagascar, south of the Bay of Bengal, and north of the Diamantina Fracture Zone. In the Indoysian Foldbelt, the seamounts are aligned E-W, and they seem to merge with those on the 90-east Ridge. On-ridge seamounts are very scarce, even on the GEOSAT structural map. The larger ones appear on a N-S azimuth on the 90-east Ridge. Two smaller chains appear to the east and are associated with the Investigator Fracture Zone. An even smaller number appear to the east of that, and they may be an extension of those at the trench segment which included the large volcanoes such as Krakatau. The Indian Ocean does not have many seamounts, possibly because the ocean basin is filled with oceanic ridges.

Truly, the fracture zones in the Indian Ocean basin seemingly invite descriptions such as "chaotic, jumbled, and confused". Five sets of lineaments exist in the Indian Ocean basin. Obviously some, some of these must cross each other, especially the trans basinal ones.

At this time no fracture zones have been sampled for age to my knowledge.

3. Pacific Ocean Basin

Only two of the Pacific basin fracture zones are shown here. The bathymetry (Figure 15) and GEOSAT (Figure 16) are sufficient to make a complete structural discussion of this basin. We will outline probably the most well-known of these, the Mendocino Fracture Zone. In addition, we will outline a N-S fracture which contains many well-known segments but is not usually thought of as a fracture zone, primarily because of the plate tectonic definition of fracture zones.

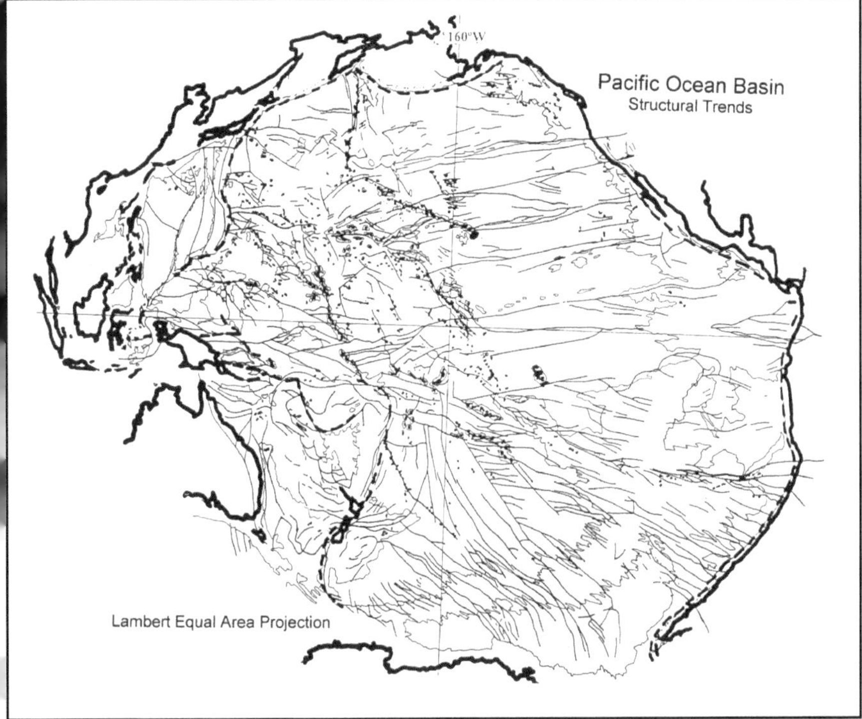

Figure 16. 1995 high-pass filtered GEOSAT structural diagram of the Pacific basin. In this diagram the trends take on the appearance of flowing eastward, with much splaying at the ends of the features. Most of the major trends appear to originate at the western Pacific trenches (dashed lines), with the regions in the western marginal seas taking on a different gross morphology.

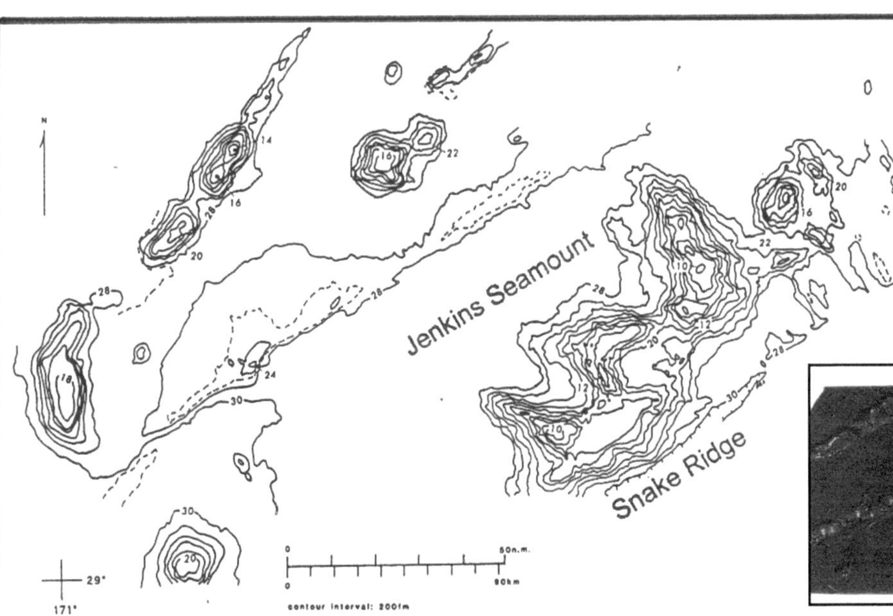

Figure 17. Total-coverage multibeamsonarbathymetry of the western extension of the Mendocino Fracture Zone double trace at a 200-fm contour interval. The Snake Ridge bathymetry has all the appearance of a linear ridge after having been pushed together by compression forces.

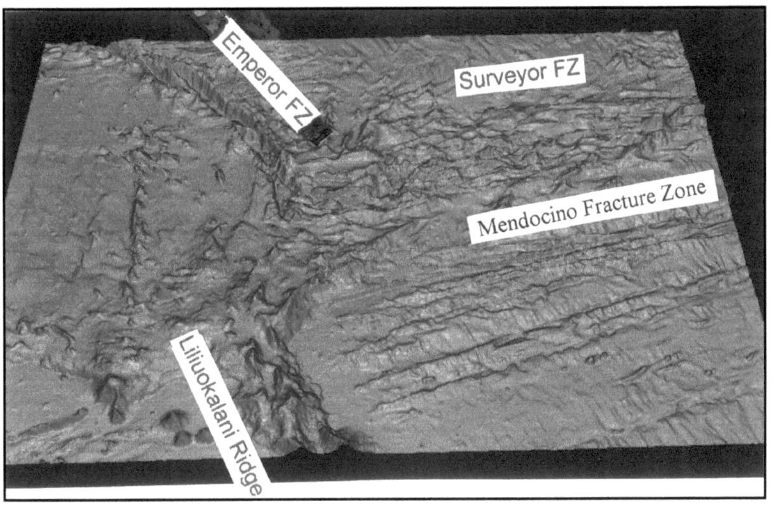

Figure 18. Regional 3D of an intersection where the Surveyor and Mendocino fracture zones intersecttheKrusenstern/Emperor Megatrend. The Emperor portion, after a slight offset caused by horizontal tectonism, continues SSE as the Liliuokalani Ridge.

United States
2
15
17
11
16
3
14
8
3
Aleutians
25
30
3
13
28
19
29
24
6
7
5
21
27
4
10
12
Kamchatka
9
1
20
New Guinea
Marianas

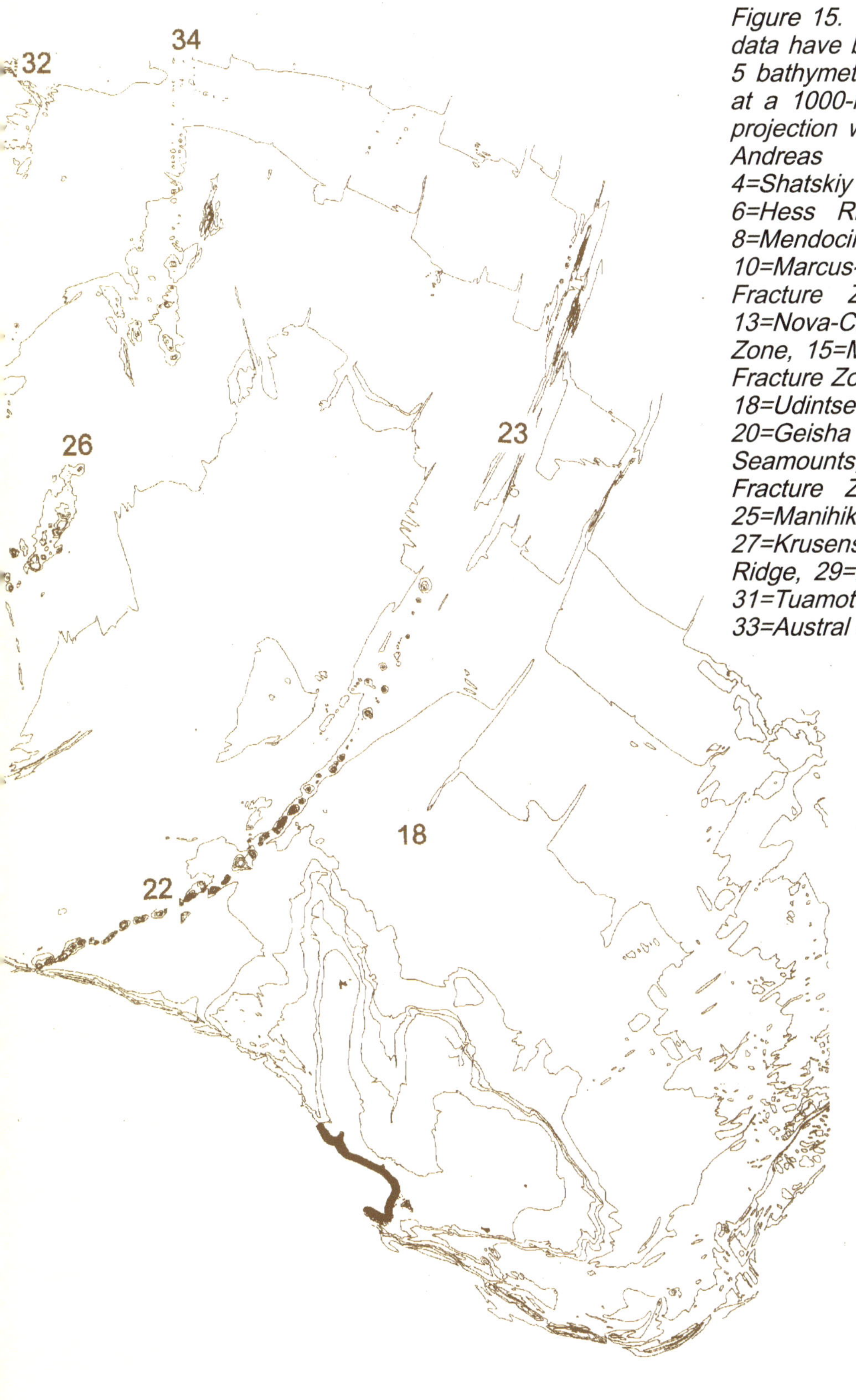

Figure 15. Where possible, multibeam sonar data have been incorporated into this DBDB-5 bathymetry of the proposed Pacific "plate" at a 1000-m contour interval on a Mercator projection where: 1=Michelson Ridge, 2=San Andreas Fault, 3=Chinook Megatrend, 4=Shatskiy Rise, 5=Emperor Seamounts, 6=Hess Rise, 7=Emperor Fracture Zone, 8=Mendocino Fracture Zone, 9=Dutton Ridge, 10=Marcus-Wake Seamounts, 11=Clipperton Fracture Zone, 12=Ontong-Java Plateau, 13=Nova-Canton Trough, 14=Murray Fracture Zone, 15=Molokai Fracture Zone, 16=Clarion Fracture Zone, 17=Galapagos Fracture Zone, 18=Udintsev Fracture Zone, 19=Fiji Plateau, 20=Geisha Guyots, 21=Marshall-Gilbert Seamounts, 22=Louisville Ridge, 23=Eltanin Fracture Zone, 24=Mid-Pacific Mountains, 25=Manihiki Plateau, 26=Tubuai Ridge, 27=Krusenstern Fracture Zone, 28=Liliuokaliani Ridge, 29=Hawaiian Ridge, 30=Line Islands, 31=Tuamotu Ridge, 32=Sala y Gomez Ridge, 33=Austral Seamounts, 34=Chile Rise

The Mendocino Fracture Zone (Figure 15-8) begins on the western Pacific forearc, crosses the trench through the Dutton Ridge (Figure 15-9) and continues through the Marcus-Wake Seamounts (Figure 15-10; discussed later). From there it appears as a double trace in the form of the Snake Ridge (Figure 17) and the northeastern double trace. Continuing through the southern Emperor Seamounts, the fracture skirts the south end of the Hess Rise and continues across the rest of the eastern Pacific basin as the Mendocino Fracture Zone (Figure 18), a feature which goes ashore all the way to the Yellowstone National Park. The eastern portion seems to form a rhomboid structure and then splay, encompassing the Surveyor and Pioneer fractures zones in both data sets.

And, this exercise could be performed for the Murray (Figure 15-14), Molokai (Figure 15-15), Clipperton (Figure 15-11), Clarion (Figure 15-16), and Galapagos (Figure 15-17) fractures too, with the first three seemingly to splay from a common point at the Ontong-Java Plateau.

The Krusenstern Fracture Zone is the one N-S feature to be discussed here (Figures 15-26 and 19). Beginning to the west of the Obruchev Rise at the Kuril Trench, this double-ridged feature becomes a single, downdropped fracture as it passes into the Emperor Seamounts (Figure 15-5). It manifests itself again on the western boundary of the Musician's Seamounts as the Liliuokalani Ridge (Figure 15-28), passes through the Hawaiian Ridge (Figure 15-29), and continues southerly as the Line Islands (Figure 15-30) and the Tuamotu Ridge (Figure 15-31), where it bathymetrically stops somewhere in those sediments.

The fractures zones, as home to many seamounts, will be discussed after the next chapter.

SEAMOUNTS, ISLANDS, and GUYOTS

The presence of "bumps" on the ocean floor, disregarding islands, was discovered in 1889 by J. Murray. The bumps, which were seamounts, were contoured essentially as concentric circles until multibeam sonars were introduced in the 1960s. Accepted use has placed seamounts in the category of volcanoes.

Seamounts are the most numerous of the ocean floor features, although some would have them to be too numerous. Seamounts start as knolls, evolve into hills, and then to seamounts. If they continue to grow into the subaerial region, they become islands. These features are the result of primary geomorphological forces in action. Secondary forces take over in that regime that tend to reduce these features to sealevel, where they become atolls if they lie in the carbonate zone. Further subsidence produces guyots once they sink to below 200m.

A study of near-total coverage bathymetry in the North Atlantic gives the seamount location for that basin (Figure 10). The size-frequency distribution is such that the smallest volcanoes are much more numerous than the larger. The reader must bear in mind that, while the number of black smokers, knolls, and hills increase the number exponentially, they are not seamounts by definition and are not included in the count, that about 900 seamounts for the entire study area.

Seamount formation is of two types. On-ridge seamounts are formed at the crest of the MOR and sometimes continue to grow as they subside. In the Atlantic on-ridge seamounts exist primarily from the Hayes Fracture Zone to the north, including Iceland. On the basis of morphology and structural evidence, most small seamounts probably originated near ridge-transform intersections and fracture zones and are round, although the shape is thought to be determined by hydraulic resistance to the flow of magma through the edifices. Their life-span is very ephemeral for the most part.

Figure 10 shows that on-ridge formed seamounts are a rarity in the North Atlantic, so rare that almost none exist in nature. On-ridge seamounts actually are the famous black smokers, worm tubes, etc. that are in the popular press. According to the total coverage surveys, and that was all of the Navy's multibeam data over the years until 1998, no on-ridge seamounts are to be found with the exception of Iceland and Ascension Island in the South Atlantic. One must remember, though, that most people are constrained by definition to describe seamounts as "elevations rising more than 1000-m and of limited extent across the summit;" 50-m hills are not seamounts.

A second type, off-ridge seamounts, is generally formed by fracture zone leakage.

Linear chains were first noted by James Dwight Dana in 1838 when he noticed both the Hawaiian Islands and the Austral and Tamed ridges to be parallel to them. By looking at the amount of erosion, Dana judged the Hawaiian Islands to be older the further away they were from the Big Island. Linear chains were suspected to form along propagating fractures by volcanism. Then these linear island chains and aseismic ridges were found to form as the ocean crust moved over stationary magma sources, which were named hotspots fed by deep mantle plumes. Fracture zone seamounts have any shape.

Seamount growth has been thoroughly analyzed. Excess magmatism localizes to form low, small domes. Continual intrusive magmatism builds the edifice to 450-1800 m leading to an unfavorable hydrostatic environment. Satellite, or parasite, cones form at the base or flanks of the edifice which drain the magma system. This leads to summit collapse, caldera formation, and the institution of circumferential feeders and ring feeder dikes to start building again. The upper slopes represent the angle of repose for the complex assemblage of pillow lavas, talus, and coarse pyroclastic debris, so the slopes are in a constant state of change with the lower slopes, or aprons, growing lower angled. Such micro-relief can be detected by side-scan sonar or bottom photography. Erosion of the volcaniclastic materials, shallow water bioclastic debris, and pelagic sediment aids in this transformation. The seamount can remain in this domain for its entire existence.

Volcano height is then limited by angle of repose and lithosphere thickness, assuming the seamount is not cut off from the magma source. Seamounts are formed in approximately equal proportion by intrusive and extrusive processes. The nature of extrusive processes such as eruptive clastic and depositional mechanisms changes during the growth of a seamount if the critical depth is passed, and then shallow intrusives form flank rift zones. Flank rift zones can be thought of as rootless, two-dimensional shield volcanoes similar to Kilauea's East Rift Zone where eruptions draining the magma reservoir have built such an edifice. Intrusives form the earliest feeder dikes and sill swarms, and extrusives form layered beds or lava ponds.

Island building is the another step in a seamount's evolution. Once the seamount breaks the surface of the water, it becomes an island. The effects of secondary erosion such as mass wasting and stream erosion quickly reduce these islands to banks, which are just below the surface of the water. From the bank stage, a seamount will subside enough to reach the depth reserved by the USBGN for guyots.

During World War II sonar had progressed enough for flat tops to be noticed on some Pacific Ocean seamounts by then-Navy man Harry Hess. A study revealed that these features varied in size, had flat or gently sloping tops of 2° or less, and were circular or oval in plan. Hess named them guyots after a 19th century geographer, Arnold Guyot. The observed top depths of the flat surfaces ranged from 520-960 fms and the bottom from 2600-3100 fms. The guyots as described by Hess exhibited very little terracing and were thought to be relics of Precambrian volcanic islands with no reef growth. The sonar was a minimum 65° beam width, and navigation was poor by today's standards.

Edwin Hamilton (1956) showed several guyots in the Mid-Pacific Mountains to be submerged between 700-900 fms. These features displayed symmetry and concave sides upward to 20° slopes near the tops. Conclusions from this study were as follows: Guyots of the Mid-Pacific Mountains were truncated volcanoes with flat, wave-eroded platforms; reefs tried to become established during the Cretaceous but were killed by rapid submergence; and the guyots continued to subside to a 900 fm summit depth. The evidence then available did not support theories of sediment-filled calderas and atoll lagoons, or upfaulted blocks. Here again, the sonar beam width was 65°, and the navigation was marginal to poor by today's standards.

The "state of the art" was restricted to the definition of a guyot as a flat-topped seamount deeper than 100 fms (200 m). They were not thought to be drowned atolls. The topography was the same as that of an insular shelf except for island erosion. None of the dredged guyots displayed anything older than Cretaceous material on them. The depths to the summit ranged from 100 to 250 fm, with most falling between 500 to 1000 fm. Guyots were felt to be worldwide, with half of the estimated total of a few hundred lying in the West Central Pacific. One important note was that the depth should be considered as the break in slope between the sides and the summit plateau, because this represented sealevel when truncation began. This point has since been named the summit plateau break depth. Furthermore, the present guyot depth represents the sum of sealevel change and subsidence since formation. Guyots are said to be dipsticks to record ancient ocean depths, and that the subsidence implied by guyots is related to the subsidence of the oceanic crust itself.

After a complete study of northern hemisphere guyots, the definition of guyots was found to be wanting. Hess originally placed their number at 160 between Hawaii and the Marianas. His number was increased by about 100 in the western central Pacific by Menard and Ladd in 1963. While this figure includes atolls and submerged atolls, it generally reflects thoughts about guyots at that time. Heezen and Hollister perpetuated and even expanded Menard's number in 1974. None of these were even remotely accurate. In science, seeing what one wants to see is fairly easy. A downsizing of the numbers of guyots came with the introduction of the more refined surveying methods. The resurveying of Fieberling Guyot (Seamount) and Erben Guyot (Seamount) pointed out the missed peaks and smoothed bathymetry that had been so prevalent with the older surveying methods. Another method is to grow a submerged flat-top without ever breaking the surface.

All the previous publishers of guyot mythology believed that wave truncation and subsidence were the primary forces to flatten the tops. An update of the Hawaiian and Canary ridge bathymetry used all the available bathymetry and topography. The digitized data were processed through the GRASS-GIS package. On closer inspection of guyots in the making, subaerial mass wasting and erosion would appear to be the primary levelers rather than wave leveling.

As the seamount ages, another form of erosion begins to sweep the lower flanks clear of deposition, and these figures rise steadily for the remainder of the volcano's existence. At Jimmu Seamount, 5180 km from Loihi, the lower flank slopes have increased to 13%. This increase is caused by bottom currents and mass transport processes where the sediment is moved off into the basins between the Hess and Shatskiy Rises.

The summit plateau break depth and summit present one last item of interest, the different morphologies. Some guyot tops are domed, and some are flat. A look at the Pacific guyots shows that most in the western side are domed. All of the younger guyots are flat. The Hawaiian and Emperor chains were used to determine where this change occurs. The entire Hawaiian chain through Pearl and Hermes Reefs and Midway Island is flat. At 43 Ma, Kammu Seamount and Yuryaku Seamount are domed. Several more flat tops occur in the central portion, and the rest of the Emperor Seamounts are domed. Flat or domed appears to be a function related to the continuing presence of a limited heat source at the leaky point over which the feature initially formed. Alternatively, it could be due to the cooling and compressional collapse of the circular magma vent. Nintoku and Suiko are both domed; both lie atop fractures.

A discussion of various specific seamount/island chains will introduce yet another link in the discussion to come:

1. Geisha Guyots

The first set of linear seamount/island chains in the north Pacific to be presented is the Japanese Seamounts, or more commonly called the Geisha Guyots (Figure 19). This chain is on an azimuth of 300°, or WNW-ESE. The history of the Japanese Seamounts represent the third possibility in seamount formation, one which appears as one type but is actually another.

The Geisha Guyots are in the NW Pacific Ocean basin (Figure 15-20). Because of the characteristic geomorphology, Kashima Guyot (Figure 47) may belong to the same edifice building regime that built the rest of the megatrend. It has the same steep upper slopes. Kashima is subducting into the Japan Trench at 35°50'N latitude and 142°40'E longitude and is normal faulted downward to fill the trench axis with rubble. The easternmost faults are really FRZs. Kashima is dated at Middle Cretaceous (Aptian-Cenomanian) based on limestone collected in dredge samples. The limestone, however, is part of the Cretaceous reef atop Kashima's surface-approaching paleo-level. It is not indicative of the age of Kashima Guyot itself, and certainly not indicative of the ocean floor on which this feature

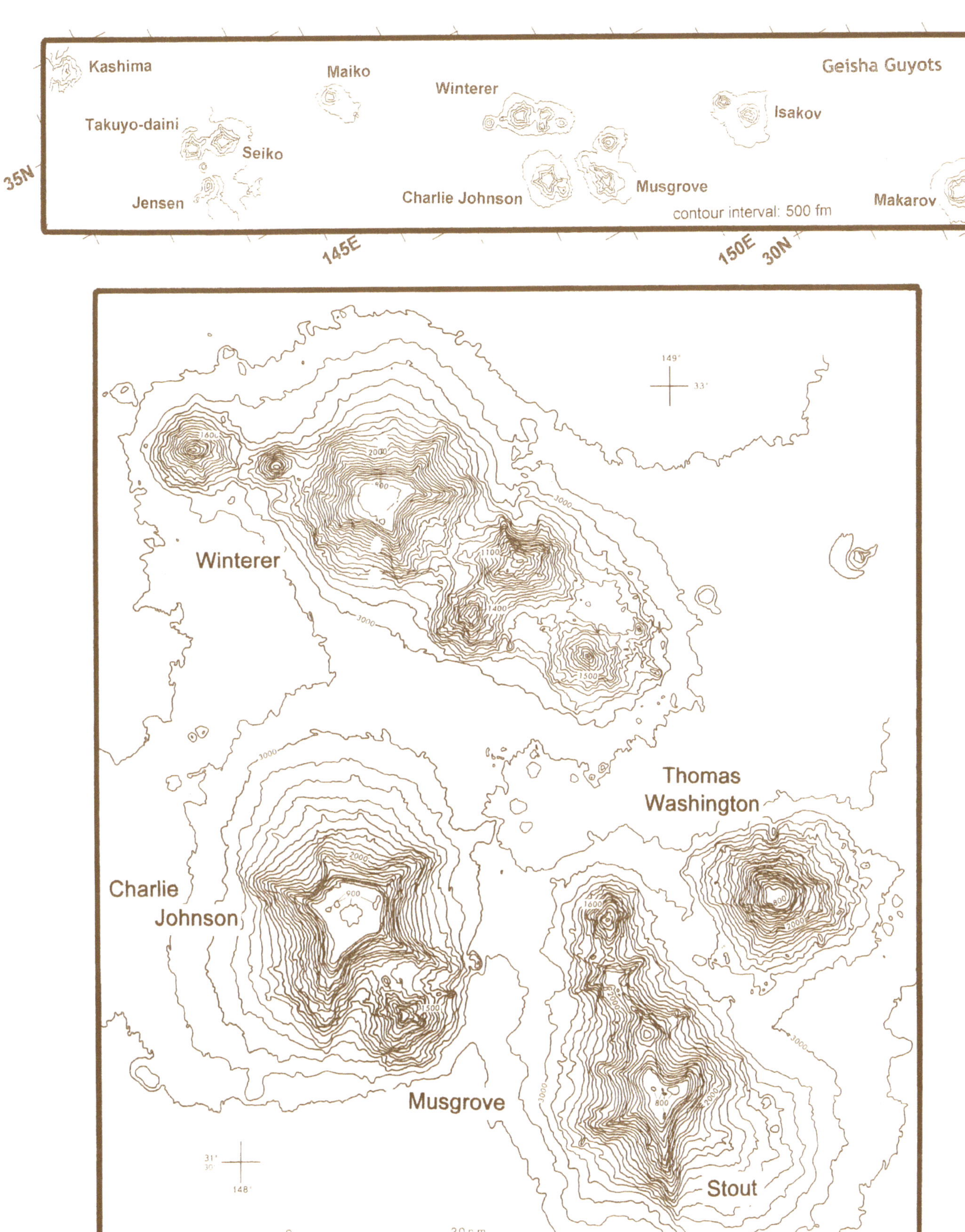

Figure 19. Geisha Guyots locator from multibeam sonar surveys. The bathymetry of the mid-chain is from NAVOCEANO's multibeam sonar data base at a 100-fm contour interval. The excessive amount of FRZs suggest an older age for these features, and that is the case. The Kashima megatrend passes through the western extreme. (named after NAVOCEANO surveyors Charlie Johnson and John Musgrove)

was emplaced by presumably later volcanism. For the time being, we will classify it in terms of the Great Cretaceous Outpouring seamounts.

Continuing to the ESE, the Seiko-Takuyo-daini Seamount group has average upper slopes of 24°. Maiko is a small seamount in the chain. The Winterer-Charlie Johnson-Thomas Washington-Stout Seamount group (Figure 19) averages around 20°, as does the Isakov Seamount group. Makarov Seamount averages 17°. All of the other measured Pacific guyots average about 13° upper slopes, and the East Pacific Rise seamounts average 18°. The second most significant fact is the comparative size of the FRZs. These are identified by the fact that they go all the way to the summit. Slumps leaving scarps that look like FRZs may only go half way up the edifice. A study of these and other FRZs shows that typically three to six FRZs appear on each seamount independent of edifice size and the average length increases with summit plateau area and pressure differential. In sequence Takuyo-Daini is 108.8 Ma, Seiko is 101.8 Ma, Winterer is 108.3 Ma, Charlie Johnson is 99 Ma, Isakov is 103.7 Ma, and Makarov is 93.9 Ma.

Maiko Seamount gives an initial clue about the growth patternoftheJapanese Seamounts. Lying at 34°02'N latitude and 145°55'E longitude between Seiko and Winterer guyots, Maiko averages 34% slopes up and down to an 1830 m summit from the regional base depth of 5500 m. It also shows two positive FRZs and two incipient ones.

Single-topped, round guyots are theoretically normal in nature, but any morphology may be started from the formational stage on, with FRZs found on 40 fm seamounts on the ridge-crest. The FRZs can give them multiple shapes such as later intrusion after formation or turbidite scouring of feature after wave action much like radial stream pattern eroding parts to leave star pattern, such as Stout Seamount and Charlie Johnson Guyot. On these features presumably wave-cut terraces, aprons, benches, escarpments, moats, saddles, spurs, and flow patterns from the original volcanism are evident. They could be a perfect example of guyot formation below the euphotic zone.

The seamounts themselves are interesting because they have steeper than average slopes for other guyots (if they are guyots). The suggestion has been made that they have grown to the euphotic zone where coral reef formation has built continuously while the seamounts have subsided, producing the steeper sides and flat tops. They are also the only in-line group of seamounts to exhibit the same height, so they have apparently undergone no vertical tectonic activity. On a WNW-ESE trend, this seamount

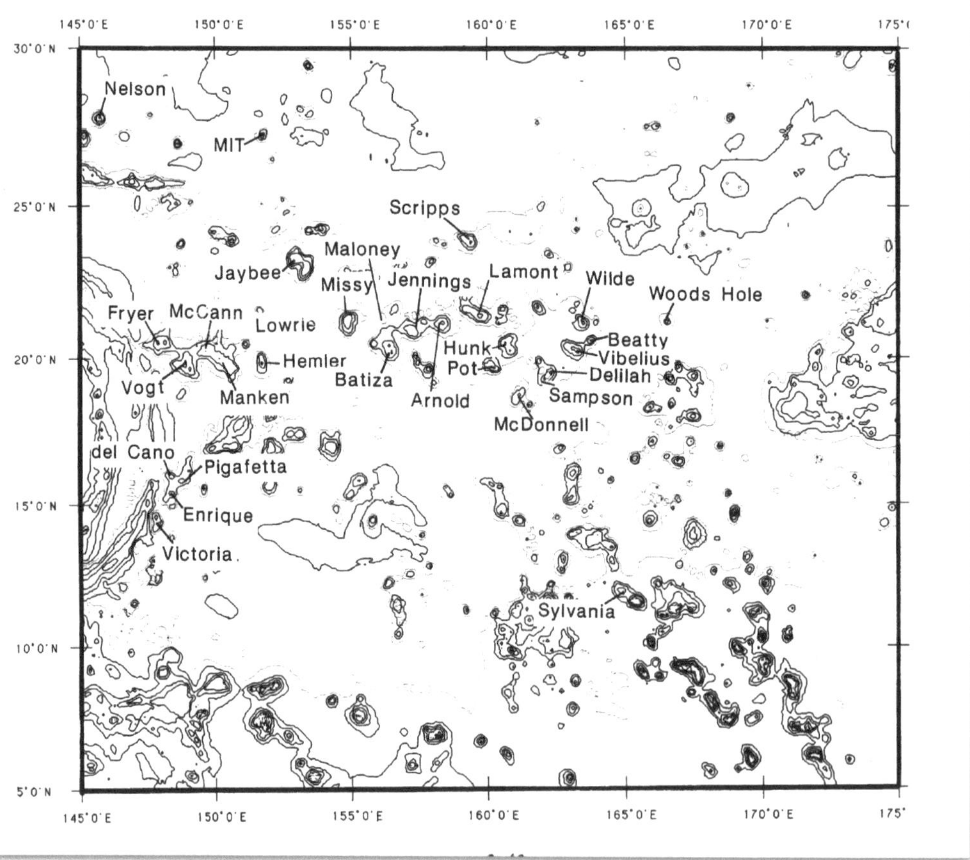

Figure 20. Locator for the Marcus-Wake seamounts from the DBDB-5..

Figure 21. The Uyeda Ridge/Nelson Guyot complex as it leaves the trench region at a 100-fm contour interval. Surveyed by multibeam sonar this ENE-trending lineament is actually the western portion of the Chinook Megatrend. Nelson Guyot is on a line with Uyeda Ridge to the NE. Several caldera remnants are still extant on Nelson's surface (named after NAVOCEANO surveyor Del Nelson and Seiya Uyeda of the Earthquake Research Institute in Tokyo).

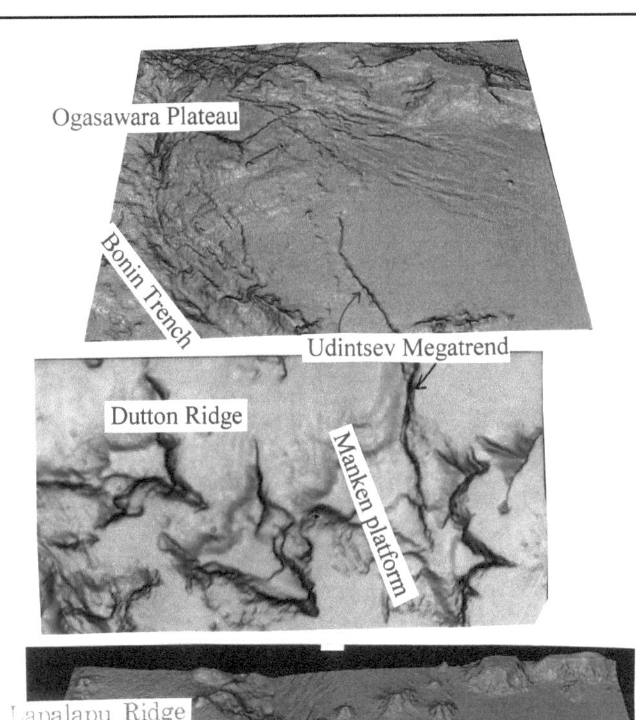

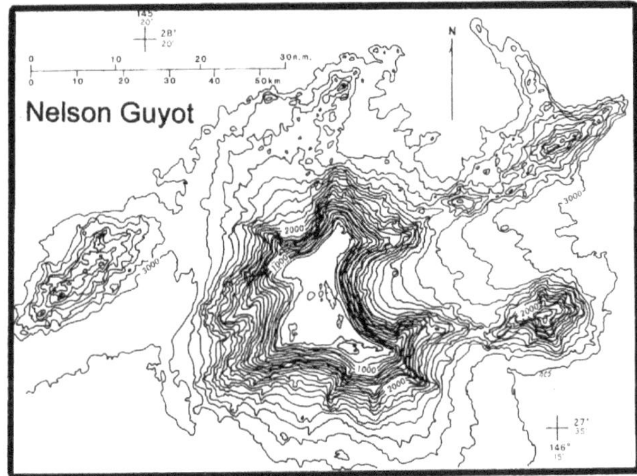

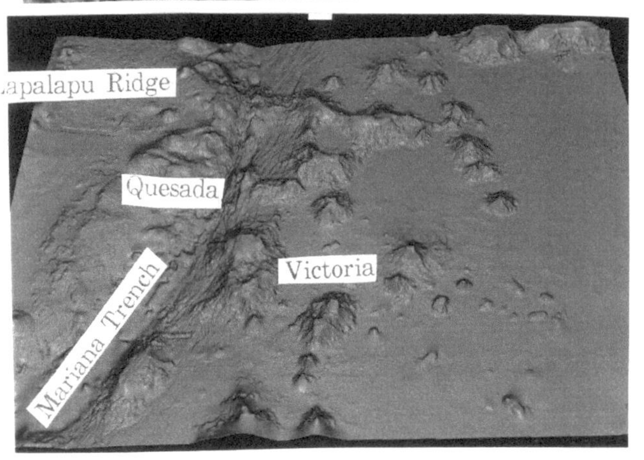

Figure 22. Multibeam sonar-based 3-Ds of the aseismic Ogasawara Plateau/Michelson Ridge (top) at the trench cusp. Fracture control is suggested for the appearance of Broken-Top Guyot. The Udintsev Megatrend begins at the plateau. The aseismic Dutton Ridge (center) from a completely vertical perspective showing the FRZs and the passage of the Udintsev Megatrend through the Manken platform. The Magellan Seamount group (bottom) at the Mariana trench. With sediment infill, the region of the upper seamounts takes the appearance of an Ogasawara sized plateau. The Udintsev Megatrend appears here as the seamount group on the right. (Ridges named after NAVOCEANO survey vessels)

chain parallels other, more recently described trends for the Pacific basin.

2. Marcus-Wake Seamounts

The Marcus-Wake Seamounts begin at the Izu-Bonin trench cusp and pass southeasterly (NNW--SSE) to the south (Figures 15-10 and 20). The northwestern-most of these features is Nelson Guyot (Figure 21) at 27°50'N latitude and 145°45'E longitude. After forming by fracture leakage, appears to have undergone later tectonic activity to show the remnants of at least five calderas.

The Michelson Ridge in the NW Pacific is presently attached to the Ogasawara Plateau (Figure 22 and book cover). The ridge is comprised of four guyots: Broken-Top, Smoot (Yabe), Castor, and Pollux.

The Ogasawara Plateau and Broken-Top Guyot head the Michelson Ridge (see cover diagrams). A fracture separates the ridge from the Ogasawara Plateau. Tilted platform tops are found on guyots, including both the plateau and the guyot at 26°05'N latitude and 145°20'E longitude. The Ogasawara Plateau is a prime example of evolving discussion on a feature's evolution. The updated views show the effects of secondary and tertiary geomorphology. The fracture bisecting Broken-Top Guyot is plainly passing south of Smoot Guyot and disappearing under the sediment. Erosion channels on the plateau between Broken-Top on the north and the large horst on the south can be followed for over 140 km. These channels were apparently the agents for transport for most of the materials created by subaerial erosion and appear to be still active. Primary tectonic forces built the plateau and fractured it. Secondary geomorphic forces did the subaerial mass wasting, such as the landslides and slumps, and erosion. Tertiary geomorphic features, such as the channels, carried that material out onto the plain. This is ocean floor geomorphology in action. In a 3-D representation of the Ogasawara Plateau, several branching channels are seen to be building an outwash plain coming out of the page. The sediment being deposited by these channels has obliterated the bathymetry of the fractures bisecting Broken-Top Guyot.

Smoot (Yabe) Guyot has been dated by many dredge samples. Among the more interesting ones are reddish brown clays, foraminiferal sands, quartz fragments, tuff fragments, and a host of "critter" samples. They have been dated at Middle Cretaceous (Albian; about 105 Ma, Shiba, pers. comm. 1998). This feature has another of the tilted tops.

Secondary ridges off Castor and Pollux Guyots have not been discussed. Flank rift zones can apparently strike off in any direction much the same as on the Geisha Guyots. The FRZ between Smoot and Castor is 110-km long. The northern FRZ on Castor is about 28-km

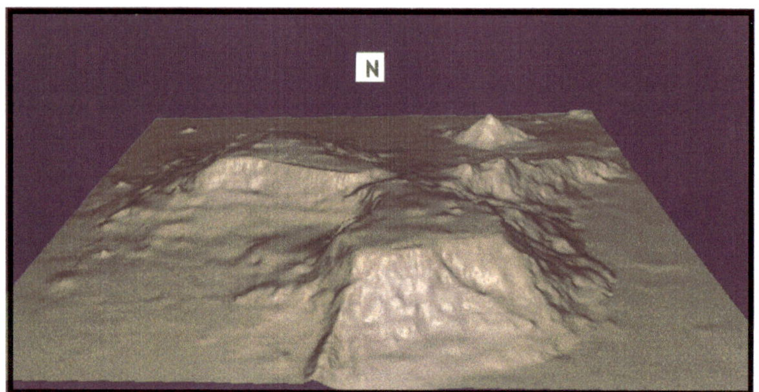

Jaybee Guyot

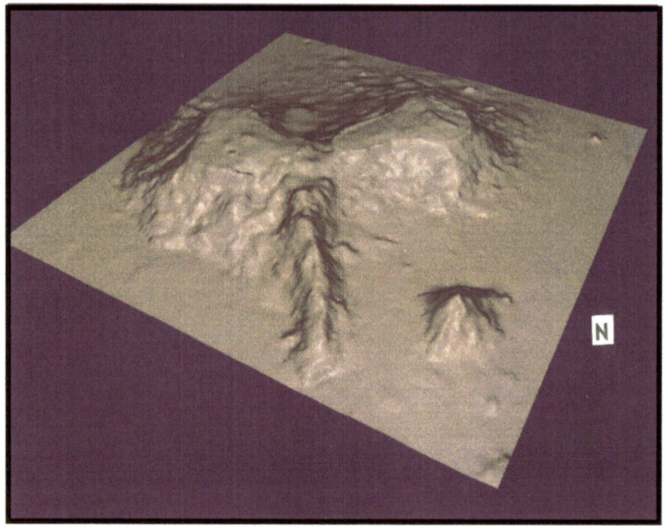

Figure 23. Multibeam-based 3D of Jaybee Guyot. The prominent FRZ on the 056° azimuth is the same as that for the Mendocino Fracture Zone, and it is in line with that feature. Did the fracture rend the single guyot asunder, or is it two distinct seamounts?

long and 26-km wide. The rough topography even at this scale shows the direction of lava flow. Pollux Guyot fits into the notch and probably contributed its share of the lava to create a much larger flank ridge. Pollux shows multiple flank rift zones itself besides its shared one. A south-trending 83-km rift lies east of that. On the eastern side is a ridge 150-km long and does not show the same tectonic fabric as the joint rift. Interestingly, it has a flattish surface and is 1100-m deeper than the main plateau on Pollux Guyot. The rift on the north, however, does show the same tectonic fabric and is one of the finest examples of its genre. Several side vents and perhaps a failed rift on the NW exist. The Michelson Ridge is 650-km long on a N85°W strike. Secondary mass wasting and erosion are evident in the back-filled "dams" on the northern flanks.

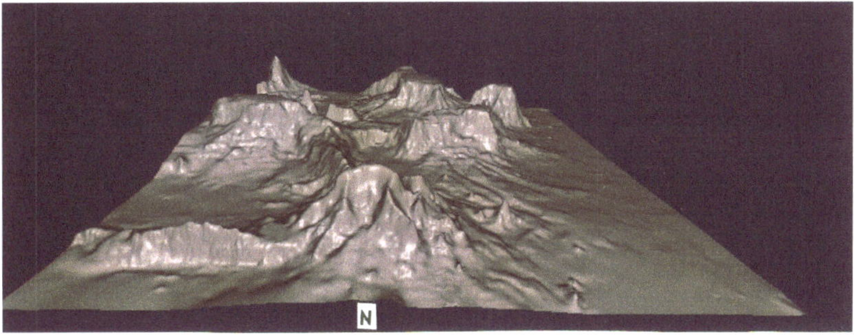

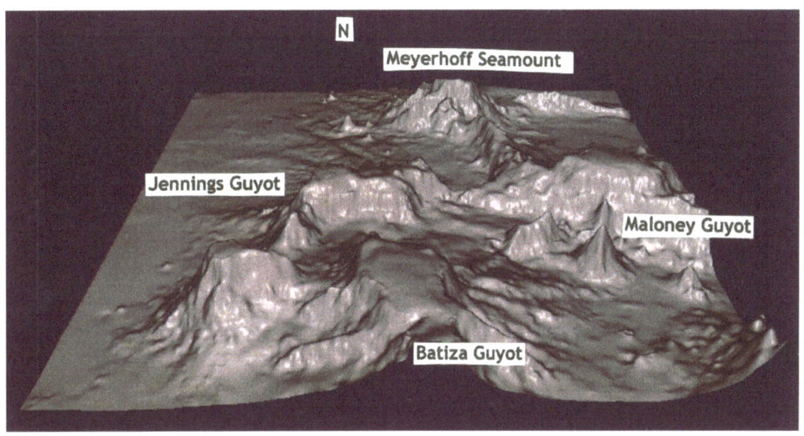

Figure 24. 3D of the cluster of guyots and seamounts in the center of the Marcus-Wakes based on multibeam sonar which could be a book in itself. Meyerhoff has a huge caldera, two possible FRZs, side vents, and an outwash plain. The question remains as to whether the FRZs are that, or are they intersecting fractures? The top view gives all the appearances of a fracture passing through the region including the vents associated with Maloney. Evidence of mass wasting and secondary geomorphologic forces at play here on all the seamounts. (Named after Jim Maloney and Chuck Jennings, early NAVOCEANO DEWLINE surveyors, Art Meyerhoff, and Rodey Batiza of HIG at the time)

Jaybee Guyot (Figure 23) appears to have formed as one edifice on an azimuth of 330° and been faulted into two by later fracturing, the azimuth of that fracture being 051°. We now know that the was caused by the Mendocino Fracture Zone. The feature lies at 23°10'N latitude and 153°20'E longitude. It has a 22% upper slope and a 4% flank slope. Rising from a regional base depth of 5500 m to a 840 m summit, Jaybee Guyot covers over 60,000 km². Was Jaybee formed by leakage into a pre-existing fracture, or was Jaybee a single edifice that was fractured later during tectogenesis of the Mendocino Fracture Zone?

The cluster of Jennings, Batiza, and Maloney guyots (Figure 24), Meyerhoff Seamount, and their attendant seamounts presents a discussion of presumably old-age seamounts. Maloney Guyot lies to the east of Jennings Guyot at 21°N latitude and 157°15'E longitude. Maloney rises on 4% lower flank slopes through 24% upper slopes to a 505 km² summit. It is on a 290° azimuth. Maloney has two attendant seamounts to the south and a tremendous flank rift zone to the north. Immediately adjacent to Maloney Guyot lies Arnold Guyot. Lying on a 275° azimuth, it rises on 4% lower flanks through 24% upper slopes to a 1465 km² summit. Meyerhoff Seamount, lying to the north, is bisected by two ridges, presumably fracture-controlled. The degree of activity is unknown at this time, but a prominent caldera faces to the north, while side vents abound on the western slopes.

Jennings, along with Maloney Guyot to the NE, is on the ENE-alignment of all the eastern Pacific fracture zones. Meyerhoff Seamount, a ridge on the south of that, and a seamount to the SE of Maloney are on the NNW-ESE strike of the great southern and western Pacific megatrends. In this region that megatrend is the Marshall-Gilbert chain. This means that the cluster has in part formed in leaky fractures at different times. The northern part of Batiza may have been sheared off to the east when the later WSW-ENE-trending fracture formed. The results of secondary FRZ formation, mass wasting, and tertiary submarine erosion are rampant. Batiza's eastern flanks and Jennings' NW flanks show a fantastic assortment of submarine channels with their attendant outwash plains. [This is the finest example of most of the tectonic and geomorphic processes in one figure in this book.]

Lamont Guyot lies at 21°30'N latitude and 160°00'E longitude. Lamont Guyot is truly one of the egregious features concerning geomorphology. Lamont Guyot shows a WNW-flow pattern off the primary structure. With most of the other guyots in this region on a more-or-less 330° azimuth, Lamont is at 295°. Lamont Guyot rises from the regional base depth on 3% lower flank slopes through 27% upper slopes to a 1280 m summit of 1570 km^2 areal extent. Lamont has seven FRZs. One could draw the conclusion that Maloney, Arnold, and Lamont guyots all formed at a later time, but they did not. A good date on Lamont is 90.5 Ma.

The final group is treated as a single entity, starting on the northern extreme. Wilde Seamount rises about 900 m on 4% lower flank slopes up through 20% slopes. Wilde was formed at 90.6 Ma. Beatty Guyot rises on 23% flank slopes to a summit at 1450 m of 404 km^2 areal extent. Vibelius Guyot rises on an 18% flank slope to a 1830 m tilted summit of 188 km^2. Sampson Guyot, at 20°50'N latitude and 163°15'E longitude, rises on 2% lower flank slopes through 22% upper flank slopes to a 1280 summit of 620 km^2 areal extent. This feature has a 2400-m parasite cone to the west. Vibelius, Wilde, and Sampson line up on a 056° azimuth. Delilah Guyot rises from the regional base depth of 5300 m on 3% lower flank slopes through 25% upper flank slopes to a 1465 m summit of 157 km^2 areal extent. The last of the guyots in this cluster is McDonnell Guyot. The pedestal rises about 1000 m over the regional base depth on 3% flanks. The 16% upper flanks rise to a 1280 m summit enclosing 450 km^2. Five other seamounts and parasite cones inhabit this pedestal. The smaller 18-km diameter seamounts each have two to four FRZs. The overall alignment of McDonnell, Sampson, Deliliah, and the attendant seamounts is 056°.

3. Magellan Seamounts

Arising at the Bonin-Mariana Trench cusp and passing WNW--ESE, the Magellan Seamounts are another of the great North Pacific seamount megatrends on this azimuth. The megatrend ends mid-basin at the juncture with the Marshall-Gilbert megatrend. We begin this discussion with the guyots of the Dutton Ridge (Figure 22).

Fryer Guyot lies at the head of the Dutton Ridge at 20°30'N latitude, 140°E longitude. It appears to have an eastward extension, as if the feature formed to the west of the trench and poured through the cusp. Fryer rises from a regional base depth of 5100 m through upper 17%-angled slopes to a flat summit at 1300 m and 1115 km^2 in extent. A deeper and smaller plateau lies at the 2560 m mark. The 3800-m tall Fryer has two basic alignments, one of 090° and the other of 308°.

Vogt Guyot lies to the southeast of Fryer at 19°50'N latitude, 149°E longitude. Vogt rises from the regional base depth at 4600 m through relatively steep upper slopes at 20% to a summit depth of about 1600 m. The 3000-m tall Vogt Guyot lies on a 308° azimuth, or WNW--ESE, just the same as Fryer. This suggests co-eval formation, or at least formation by the same mechanism. The area inside the summit plateau break depth is 2180 km^2.

McCann Guyot lies at 20°10'N latitude and 149°39'E longitude. This 5300-m tall guyot rises on lower flank slopes of 5% through upper flank slopes of 16% to a 370 km^2 summit at 1420 m. McCann trends 250°, or WNW--ESE.

Manken Guyot on the Dutton Ridge shows the effects of secondary and even tertiary platform building and erosion. Manken lies at 20°N latitude and 150°10'E longitude. The guyot rises from a regional base depth of about 5600 m to a 500 km^2 summit at 1730 m, making Manken almost 4000-m in height. Slopes of 25% on the north and 14% on the south suggest different formation regimes. The elongated plateau at 2830 m, a full kilometer lower than the summit, also reiterates this possibility. Manken Guyot is also on the basic NNW-SSE alignment of the Magellan Seamounts and appears to be a sundered portion of an older Dutton Plateau. That plateau is at the intersection of several fractures, so any type of geomorphology is possible within these constraints.

Lowrie Guyot is the easternmost feature on the Dutton Ridge at 19°40'N latitude and 150°47'E longitude. It has 25% upper slopes and Lowrie rises from a regional base depth of 5500 m to a 1445 m summit encompassing 530 km^2. At over 4000-m tall, the conclusion is drawn that all guyots are not the same height. Nor, are they all the same depth. Where so many are in such close proximity to each other, some form of vertical tectonics must play a role. That will be discussed.

Hemler Guyot is at 19°40'N latitude and 151°40'E longitude. Hemler rises from the 5500 m regional base depth on 3% lower slopes through 25% upper slopes to a 1300 m summit of about 200 km^2. The guyot is flanked by two parasite cones. This gives the feature a 300° azimuth. Hemler has been dated at 100 Ma for the surface dredge hauls.

DelCano Guyot (Figures 22 and 25), at 16°N latitude and 148°20'E longitude, is fracturing and downdropping in addition to having FRZs. DelCano has a primary axis of NNW--SSE with a secondary axis of 056°, or WSW--ENE. The guyot itself rises from the regional base depth of 5675 m to a 370 km^2 summit area at 2000 m. The upper flanks slope 23%. On the same pedestal Pigafetta Guyot lies at 15°50'N latitude and 149°E longitude. Pigafetta is elongated E--W and is 75 km X 65 km in size. Its 516 km^2 summit is 1650 m below sea level. Enrique Guyot and a southern, unnamed guyot also reside on the same pedestal. Enrique is at 15°30'N latitude and 148°30'E longitude. It rises on upper slopes averaging 35% to a 356 km^2 summit area at 1650 m. The southern guyot, herein named Espinosa, rises on 30% slopes to a 2200 m summit area of some 230 km^2.

Four seamounts on a 330° azimuth lie to the east of the previous pedestal. The two larger ones, Trinidad and Concepcion, have 28% flank slopes. Trinidad rises to a 2745 m peak, and Concepcion rises to an 1800 m peak.

Victoria Guyot, lying at the Mariana Trench at 14° 20'N latitude and 147°45'E longitude, has two plateaus inside the summit plateau break depth, which rises from a 5600-m regional base depth. The northern one is at 2560 m, and the southern one is at 2000 m. The summit plateau area is 2245 km^2, making Victoria the largest guyot in this group. At 3000-m high, Victoria has relatively uniform flank slopes with the lower being 14% and the upper being 18%.

The southernmost guyot, Serrao, is at 13°30'N latitude and 147°05'E longitude. It rises on 18% slopes to a 3900-m platform from a 6100-m trough on the eastern slope. This feature, elongated to the NE, has a major axis of about 100 km and a minor axis of about 50 km.

4. Hawaiian Chain

The Hawaiian Chain begins with Kimmei, Antoku, and Taisho Guyots and continues through Kammu Guyot (Figure 26). Kammu Guyot demonstrates several principles very clearly. It appears as a coalesced cone where two distinct volcanoes have blended together. One has a 70% sloped cliff over 1280 m in height which could have been formed by compression due to plate-flow direction change or by slumping, as witnessed by the broadened contours directly below the cliffs. These show a certain degree of flattening not seen elsewhere on the feature. Daikakuji Seamount, a guyot formed off-axis by a hot spot, is elongated on a 076° azimuth and shows several summits which have been previously eroded. The overall geomorphology shows flutes, scarps, and flank rift zones. One of the cliff scarps has 71% slopes with bathymetric evidence of slumping. Yuryaku Seamount, Toba Guyot, Kimmei Seamount, Antoku Guyot, and Taisho Guyot are all included to demonstrate the evolution of the lower flank slopes, as well as the geomorphic assortment of flutes, scarps, flank rift zones, and intrachain alignments. In this group the lower flank slopes are 5-6%, so the bottom scouring currents are already in action. Kinmei is 39.9 Ma, Daikakuji is 42.4 Ma, and Yuryaku is 43.4 Ma.

Colahan Seamount is the site where the Mendocino Fracture Zone passes through the Hawaiian Chain. Interestingly, the gross morphology seems to be unaffected by such a monumental even, leading one to speculate that the seamount formed in the pre-existing fracture.

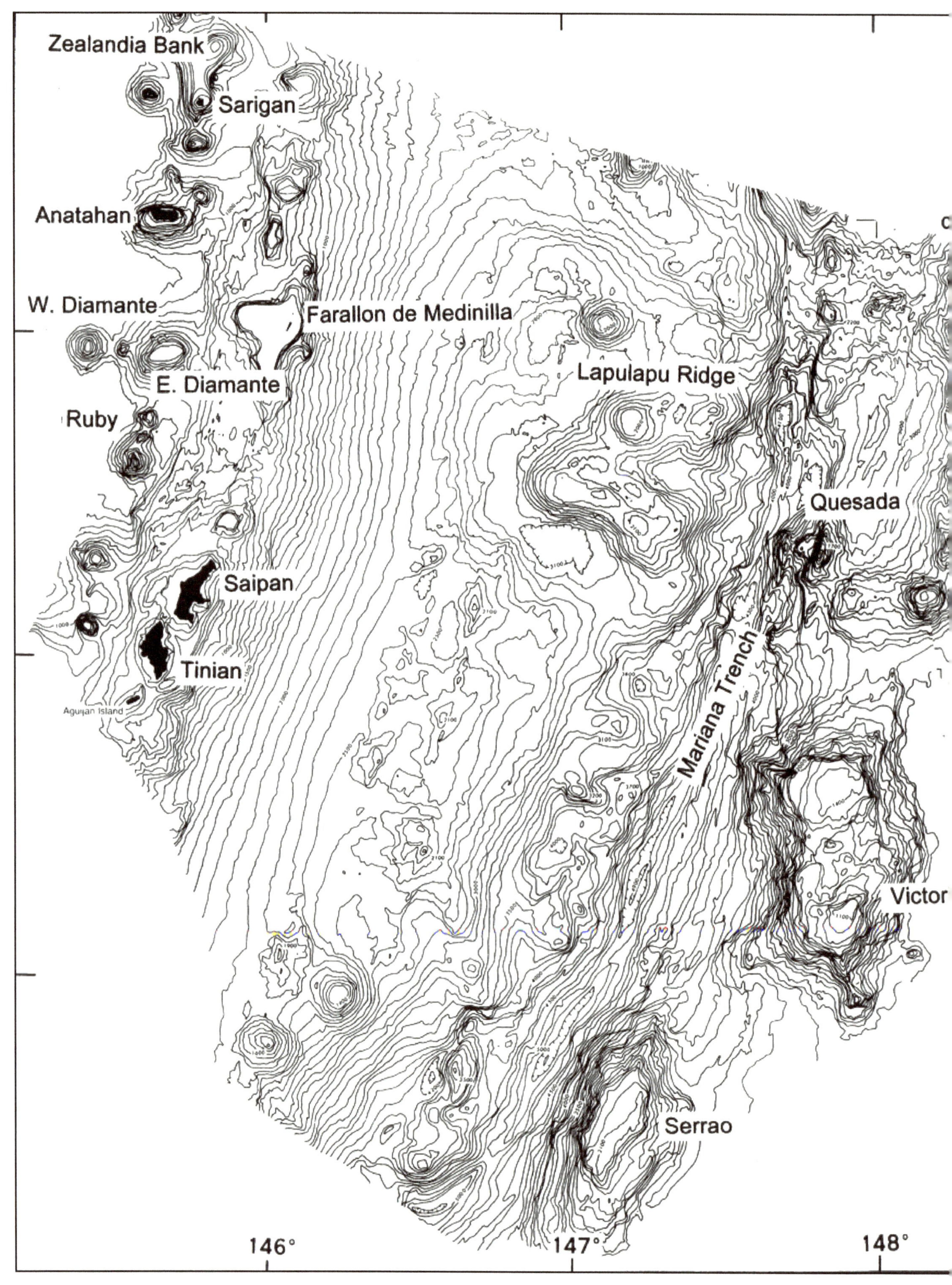
Zealandia Bank
Sarigan
Anatahan
W. Diamante
Farallon de Medinilla
E. Diamante
Lapulapu Ridge
Ruby
Quesada
Saipan
Tinian
Aguijan Island
Mariana Trench
Victor
Serrao
146°
147°
148°

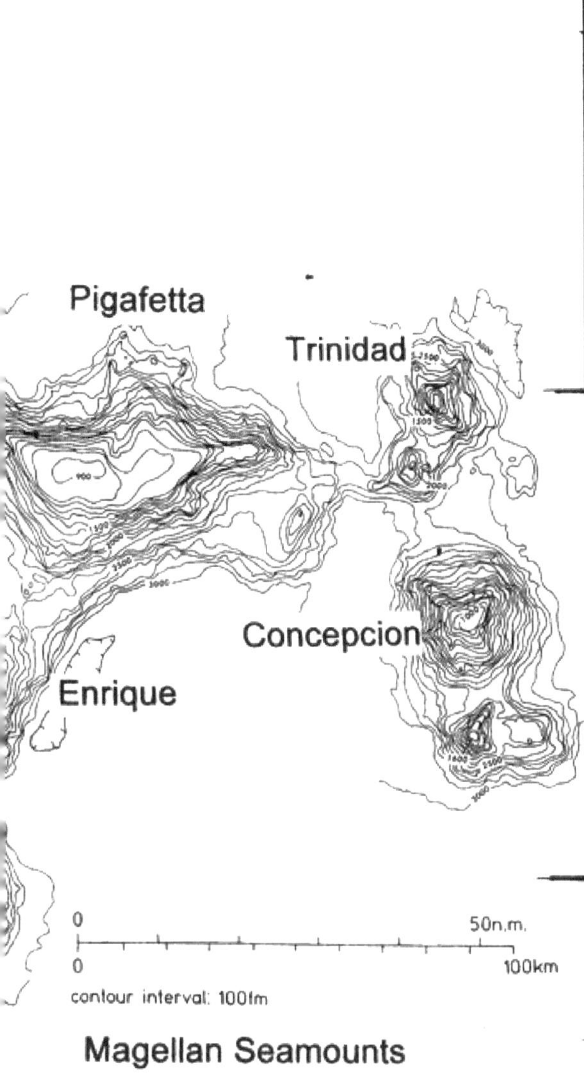

Figure 25. The Magellan Seamounts at the juncture of the Mariana trench at a 100-fm contour interval, surveyed with a multibeam sonar system. The Lapulapu Ridge could be a compression feature. This region is just top the south of a deep earthquake region, which for all purposes appear as stacked events. Some form of deep-seated tectonism is occurring here. The eastern SSE-trending seamounts are a possible extension of the Udintsev Megatrend. (All seamounts named after Magellan's crew and ships. Lapulapu killed Magellan)

The progression continues to the pedestal supporting Gardner Pinnacles, St. Rogatien Bank, and French Frigate Shoals (Figure 27). The features have become banks with the summit break depth and plateau at about 35 m. The upper slopes are 16%, which is average for the Hawaiian chain. Gardner Pinnacles is 12.3 Ma, Brooks is 13.0 Ma, and French Frigate Shoals is 12.0 Ma. Enough erosion and deposition have occurred to the surface to reduce the lower flank slopes from 4% to 1%, with most of this on the north slopes.

After the edifice is built as large as possible under given conditions, the secondary epeirogenic forces take over. Theoretically, steeper sides will grow under water rather than subaerially because they are partially buoyed by water. This is nowhere more evident than on Hawaii when the upper submarine slopes are 18%, and the subaerial slopes are 12%. Various forces such as rain, streams, wind, and mass wasting reduce the subaerial volcano to a nearly flattened surface as witnessed by the progression from Kauai through Maui. Wave terracing is responsible for undercutting such areas as the Napali cliffs on northern Kauai, but it is clearly the subaerial forces that more or less produce the flattened guyot summit before subsidence carries the surface under water. The different patterns of subsidence appear to reflect the interaction of isostatic uplift and thermal subsidence according to Menard. In uplifted regions, coral reef activity will form as fringing reefs to create lagoons and backfilling and even flatten guyot surfaces.

The Hawaiian Islands are shield volcanoes and shield volcano clusters (Figure 15-29, 26, and 28). The volcano continues to grow, sometimes forming an island such Hawaii. The East Kilauea Rift shows part of the growth pattern of a large seamount. On Hawaii five different episodes of volcanism have produced the coalesced shield; Moana Loa, Moana Kea, Kilauea, Alika, and Hualika. One can easily see the seaward extension of Kilauea volcano on Hawaii Island, the active volcano Loihi. Most investigators are excited by the prospect of yet another Hawaiian island. Loihi is nothing more than another part of the shield. Most of the magma flowing in this channel is at the eastward end, and that is where most of the activity should be expected. It is. The most apparent point is that the slopes will not build as steep subaerially as submarine. The next is that the chain of craters continues primarily down the axis of the FRZ with very few outliers. Otherwise, for the area of multibeam coverage, landform looks as would be expected.

Cross Seamount, south of Oahu, is 80 Ma.

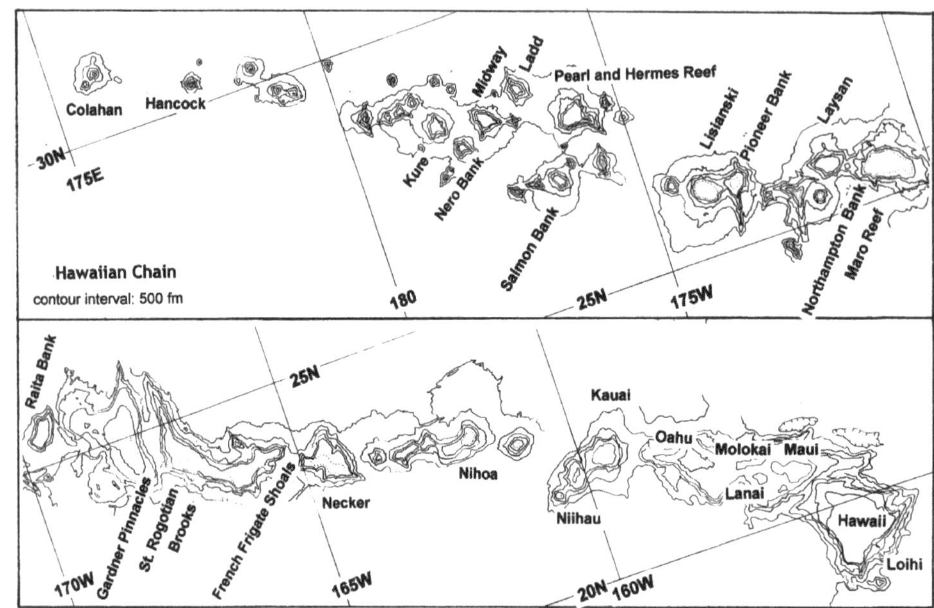

Figure 26. Bathymetry of Hawaiian chain from multibeam sonar surveys showing a wide variety of seamount-type features.

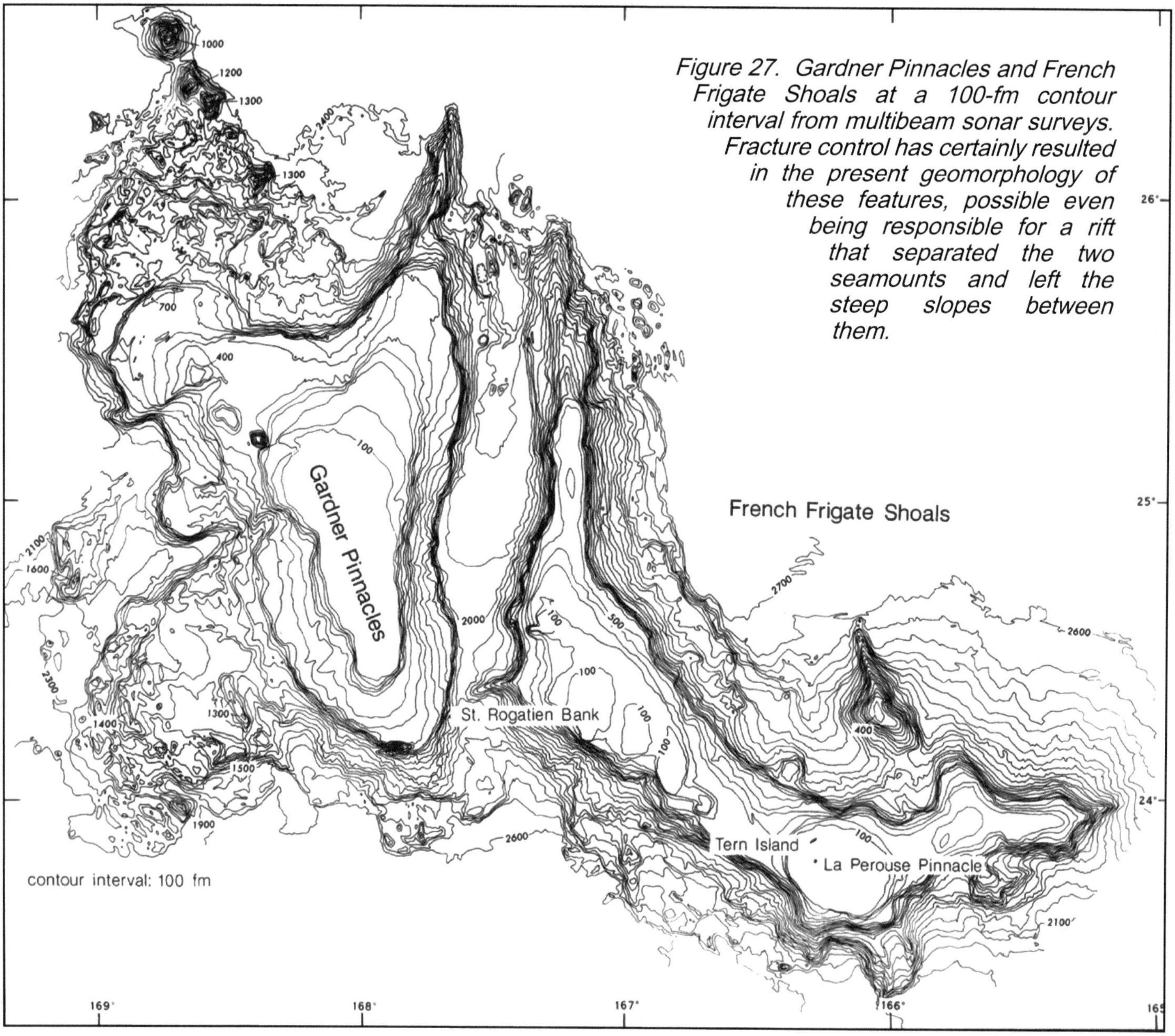

Figure 27. Gardner Pinnacles and French Frigate Shoals at a 100-fm contour interval from multibeam sonar surveys. Fracture control has certainly resulted in the present geomorphology of these features, possible even being responsible for a rift that separated the two seamounts and left the steep slopes between them.

5. Emperor Seamounts

The Emperor chain, on a NNW--SSE azimuth in the north-central Pacific, is the textbook case for linear seamount formation. This chain arises on the Kamchatka Peninsula at the active arc (Figure 29), continues across the Kuril Trench, and is part and parcel of the Obruchev Rise. The rise overprints the Emperor Fracture Zone and means that the plateau and subsequent Emperor Seamounts are younger than the Emperor Fracture Zone. The Obruchev Rise's primary axis parallels those of the Krusenstern and Stalemate fracture zones. The rise presents a 130-km front to the trench and rises 2560 m above the surrounding ocean floor on 3% slopes. The figure shows a brittle seaward lithosphere breaking up as it bends, possibly causing compression on the forearc where the Kamchatka canyons formed.

The next feature in sequence is Detroit Tablemount. The summit of Detroit has exceptional coverage. There are double coalesced cones with later volcanic activity on Detroit. Detroit, at 51°15'N latitude and 167°45'E longitude, arises from a regional base depth of 5500 m to a 1650 m peak. Detroit has been dated at Precambrian by metamorphic rocks from its surface by Boris Vasil'yev of the Pacific Oceanological Institute in Vladivostok in 1982. Fracture control is given as the reason

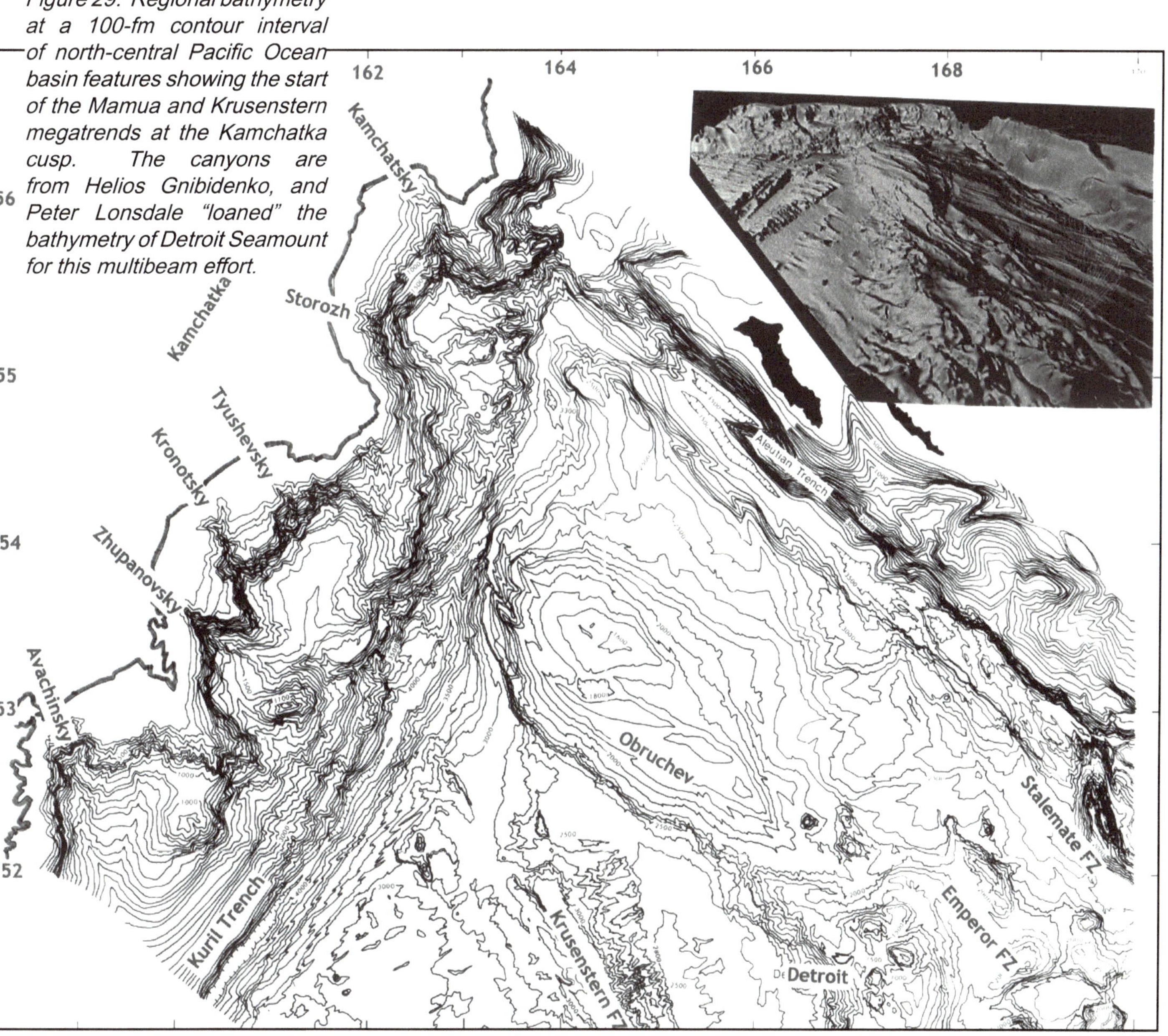

Figure 29. Regional bathymetry at a 100-fm contour interval of north-central Pacific Ocean basin features showing the start of the Mamua and Krusenstern megatrends at the Kamchatka cusp. The canyons are from Helios Gnibidenko, and Peter Lonsdale "loaned" the bathymetry of Detroit Seamount for this multibeam effort.

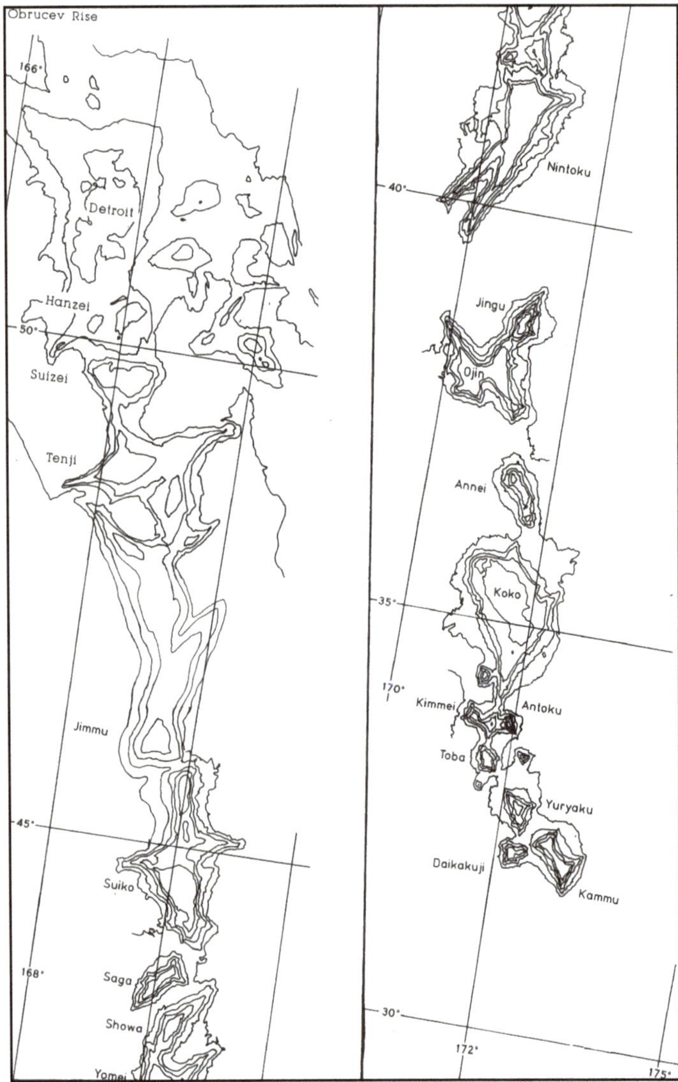

Figure 30. Emperor Seamount chain locator from various data bases at a 1000-fm contour interval.

for the north to east trends on Hanzei and Suizei seamounts. Hanzei is at 50°10'N latitude and 168°E longitude. Suizei is at 49°35'N latitude and 168°E longitude. The lower flank slopes are 14% on both the east and west, and both features are 2750 m high. Tenji Guyot has been incompletely surveyed. It appears to be a FRZ or a fracture ridge.

Suiko Seamount (Figures 30 and 31) has coalesced cones altered by fracturing. The NNW--SSE azimuth for that feature is significant because the azimuths on other features in the region are also NNW--SSE. The presence of some of the largest FRZs extant is of note. Suiko Seamount has the appearance of a hot line in that it looks like a ridge with four distinct summits, and that is the case. Suiko formed 64.7 Ma. After formation it looks as though the ridge was fractured, and the northern portion dropped several hundred fathoms by vertical tectonics. In this case the bottom currents on the east that have been scouring features to the south are blocked by the E--W rift zone.

Continuing on this volcanic pulse are Saga and Showa Guyots, Yomei Seamount, and Ninigi and Godaigo Seamounts (Figures 30 and 32). This cluster of features generally increases in size to the south. Saga Guyot is at 43°25'N latitude and 170°E longitude. Rising from the regional base depth of 5500 m to a 1280 m summit on slopes grading from 3% lower through 25% upper, Saga's surface area encompasses 485 km^2. Showa Guyot lies at 43°N latitude and 170°20'E longitude. Showa rises to a 894 km^2 summit at 1465 m. Yomei Guyot is at 42°20'N latitude and 170°20'E longitude. Its 915 m summit depth encompasses 1445 km^2.

Nintoku Seamount (Figure 31), lying at 41°N latitude and 170°30'E longitude, is a bathymetric oddity. This "molar-tooth" has the largest FRZs on any feature yet discovered, so large that the FRZs have FRZs. At an age of 56.2 Ma, the feature has been hypothesized to be a portion of the trans-Pacific Chinook Fracture Zone. Secondary geomorphological forces, such as mass wasting by destructive volcanism and caldera collapse may be responsible for the very large structures on the western flank. The radical difference in lower flank slope values, 10% on the east and 2% on the west, has two possible explanations. One is that the usual bottom currents have scoured the eastern slopes. The other is that prevailing wind, river, rain, and surface erosion which occurred when Nintoku Seamount was an island conveyed most of the sediment to the leeward side and deposited it there. This measurement could be a useful tool for paleoclimatology and physical oceanography studies. In order of formation the features in this discrete pulse of volcanic episodicity grow from the smaller Saga Guyot to the larger Nintoku on generally the same alignment of 35°. The feature gives a good idea as to the secondary and tertiary geomorphological forces in action with a host of slumps, slides, flank rift zones, and toppled features being acted upon by the subsurface water currents.

Jingu and Ojin Seamounts are triple-topped, butterfly-shaped edifices consisting of very large flank rift zones and slumps. Ojin Seamount also had enough fringing reefs to build a lagoonal

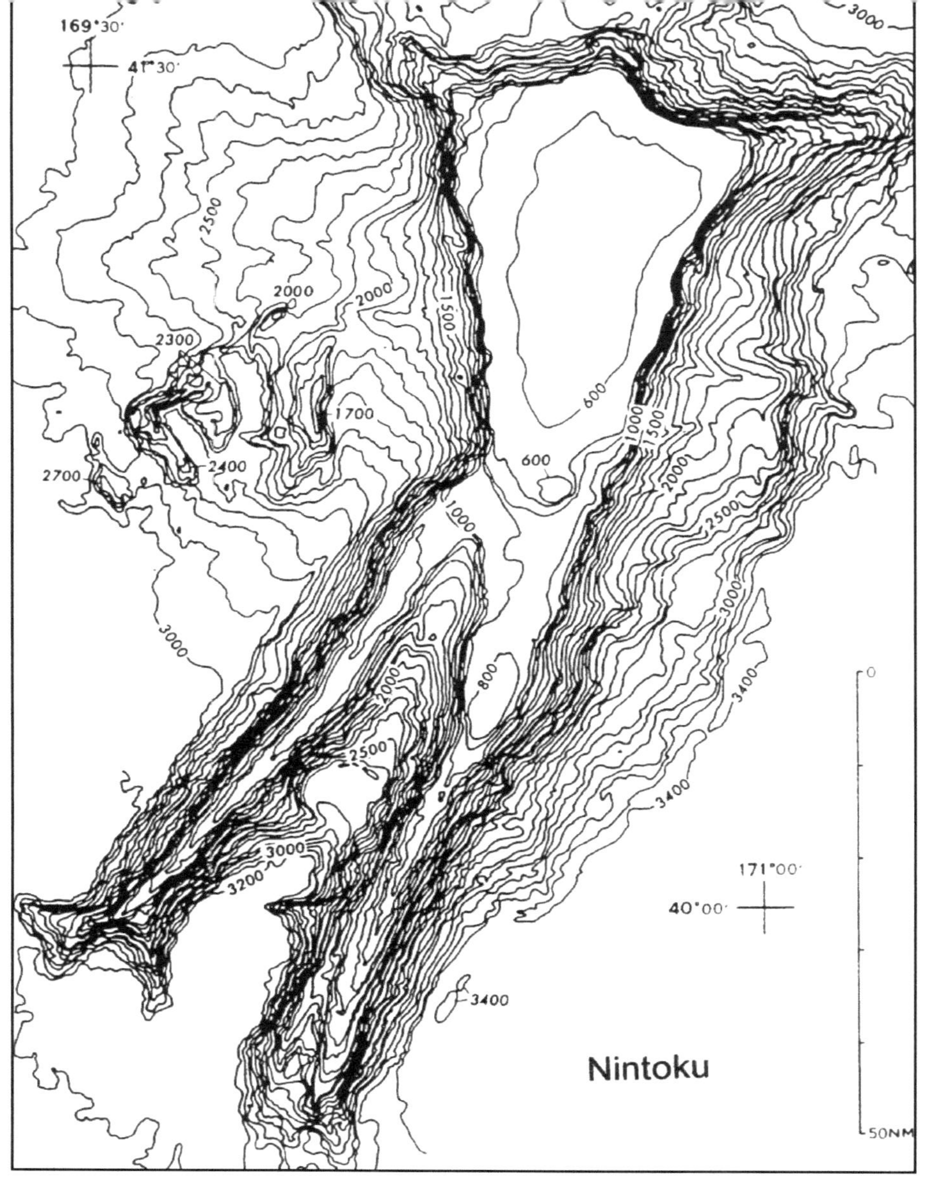

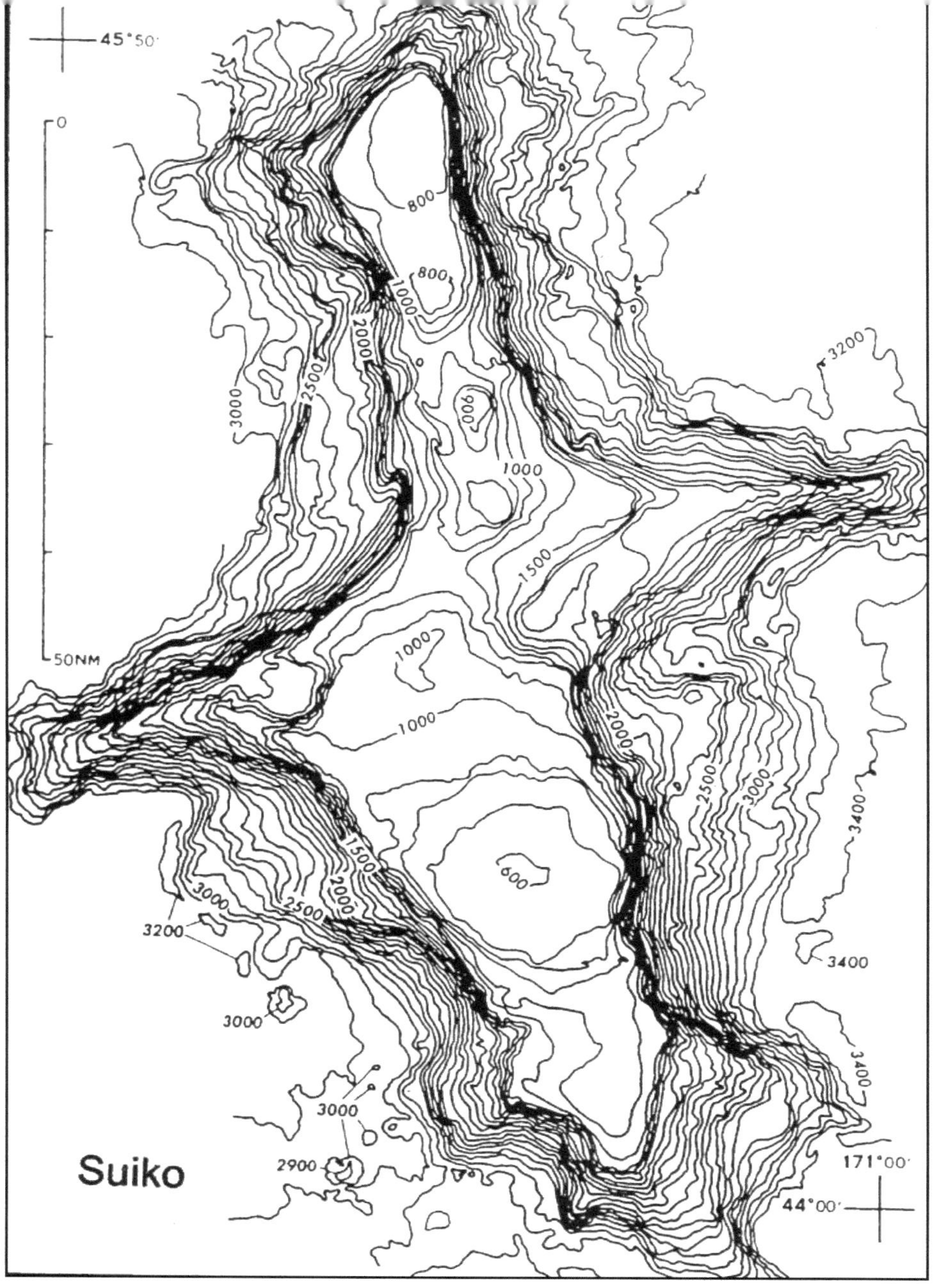

Figure 31. Suiko and Nintoku guyots at a 100-fm contour interval, surveyed totally with the SASS collector, both at same scale. Suiko (left) has undergone over 200-fm of vertical tectonism on one end. Nintoku (right) appears to have blown it's top as we see the large pebble piles on the western flank. The amount of mass wasting due to landsliding is tremendous and accounts for the low lower flank slopes on both of these very large Emperor seamounts. Fracture control is suggested for the morphology of both.

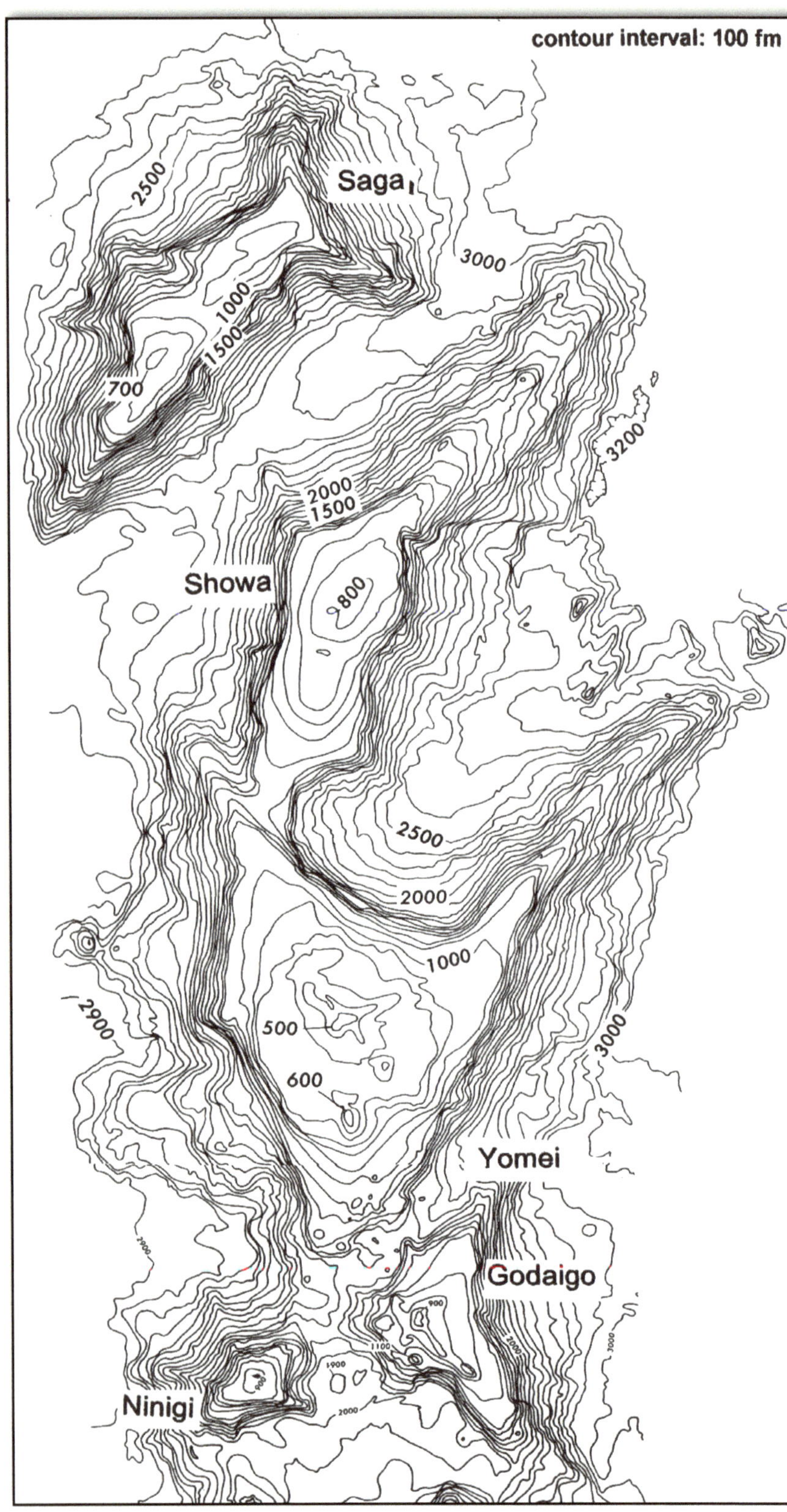

Figure 32. 1983 bathymetry of the central Emperor Seamount chain at a 100-fm contour interval from the Navy's multibeam sonar surveys. At the time, only two seamounts were thought to exist in this region as opposed to the five actually there. This was considered to be an important step forward at the time. While they lie on a N-S azimuth, fracture control is suggested for the N-E trends therein.

sediment pond so that the summit plateau break depth and summit depths approach each other in value. In its 55.2 Ma age, Ojin Guyot displays all of the features one would expect on such an old edifice. Jingu is at 38°40'N latitude and 171°10'E longitude. Ojin is at 38°N latitude and 170°30'E longitude.

Annei Seamount has a special type of morphology. On the one hand, it demonstrates how a surge channel's multiple conduit system can turn on and off in different areas to form any sort of feature imaginable. This "lazy W" is presumably in a class by itself, as no others have been found of this shape. Also of interest is that the lower flank slopes are the same on both sides. On the other hand, Annei may show fracture control due to compression. Both are possible.

Koko Seamount is a major feature at over 240 km by 180 km diameter. Koko Guyot lies at 35°N latitude and 171°30'E longitude and is 48 Ma. After the primary and secondary geomorphological forces had done their work on the feature, later primary control is again suggested in the form of fracturing. While the bottom current to the east of the Emperors has scoured the lower flank slopes to 12%, those on the west are about 1.5 to 2.0% and show either a debris field and/or parasite cones. This could be a function of a leaky fracture. Koko shows the typical rounded summit of a guyot for the most part and appears to have had at least three craters/islands.

6. Marshall-Gilbert Seamounts

Interestingly, the Marshall-Gilbert Seamounts (Figure 33) parallel the Emperors on an azimuth of about NNW--SSE, and they stop in the middle of the basin. The Marshall-Gilberts pass all the way to the Pacific-Antarctic Ridge, traverse two-thirds of the Pacific basin, and stop at the Marcus-Wake Seamount chain.

Several of the guyots and seamounts

in this chain have been surveyed, including Sylvania (Wodejebato; Figure 22), and Ruwituntun. All of these features are aligned NNW--SSE, all are on the Ralik-Ratak chains. Wodejebato lies at 12°N latitude and 165°E longitude. The summit of this guyot is 43 km long and ranges from 25 km wide on the NW to 12 km wide on the SE. It is capped by a carbonate platform.

Seamounts and islands in the Marshall-Gilbert chain are non-sequentially aged at 56--97 Ma (ODP Sites 869, 871, and 875). In fact, Sylvania Guyot has a 19 million-year range in basement ages (ODP Sites 873, 874, and 876; 65-83 Ma). Thus, the feature has undergone multiple episodes of volcanism.

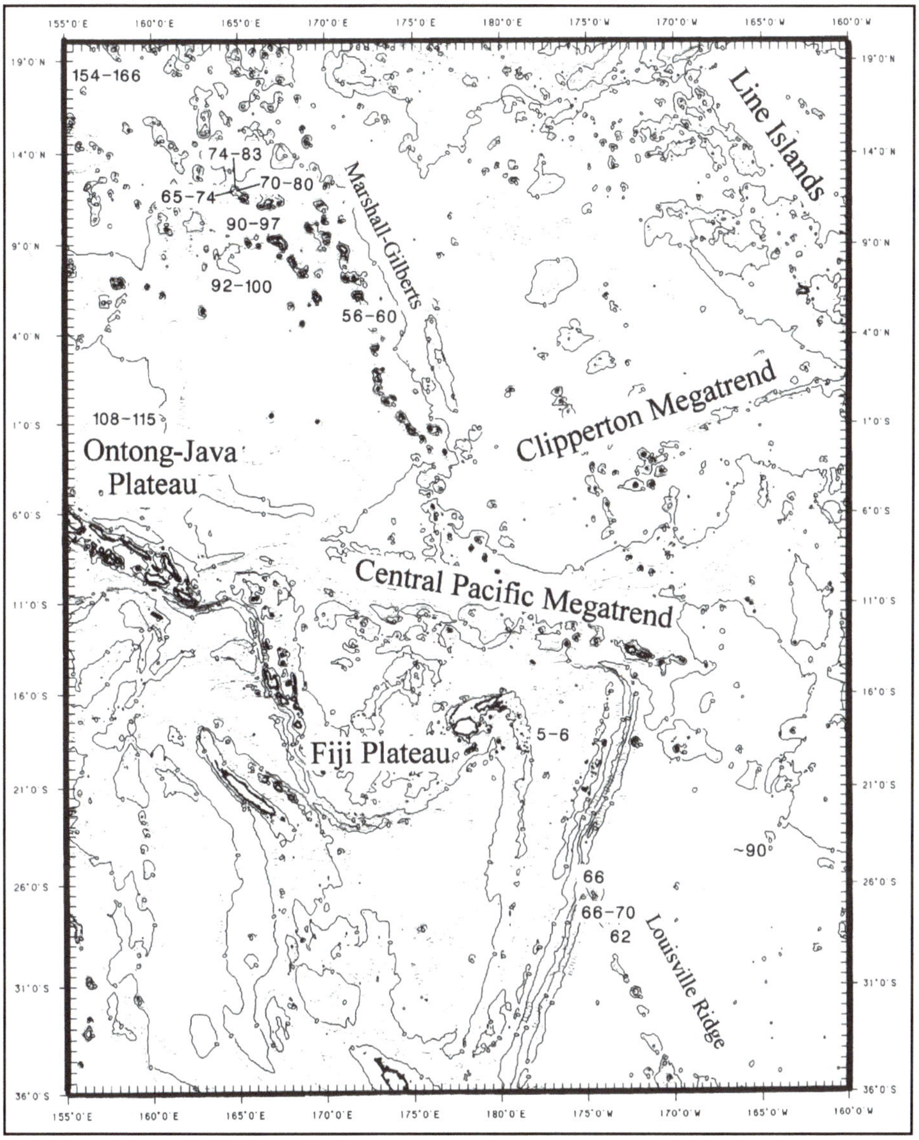

Figure 33. Locator diagram from the DBDB-5 for the central Pacific Ocean basin. While the bathymetry is not up to par for this diagram, several items are noteworthy. It is possible to see that the Marshall-Gilbert seamount chain and the Louisville Ridge could have been all the same feature before the onset of trench capture. The seamount ages are not sequential. The Clipperton Fracture Zone goes past the mythical mid-ocean boundary. And, the Fiji Plateau seems to be a huge circular feature. 1=Ralik Chain; 2=Ratak Chain; 3=North Solomon Trench; 4=New Hebrides Trench; 5=d'Entrecasteaux Zone; 6=New Caledonia; 7=New Zealand; 8=Samoa; 9=Tonga Trench; 10=Kermadec Trench; 11=Lau-Harve Basin

7. Canary Islands

The Canary Islands (Figure 34) in the NE Atlantic Ocean are a good study in guyot formation, or secondary geomorphology in action. The formation is originally attributed to wave cutting of subaerial platforms, but that has been updated. Whether or not the chain formed as the result of hotspot activity or fracture leakage is open to conjecture.

The first two peaks, La Palma and Hierro, still show craters and equal subaerial and submarine slopes. That fact disproves the idea of lower subaerial slopes. There is some debris on the lower slopes, especially below the crater breach on La Palma. Gomera is the next in-chain feature. It is already eroded to a mere shadow of its former size. The next peak, Tenerife, is variously listed at 15.7 Ma and 11.6 Ma. It is also active. Based on Gomera's erosion and Tenerife's older, more active peaks, the creation by a hotspot is now removed from the theory of the formation of this chain. The next island, Gran Canaria, is 130 nmi and 16.1 million years from La Palma. The island is incised by large canyons, and the lower flank slopes are covered by debris.

Islands on the eastern Canary Ridge, Lanzarote and Fuerteventura, are actually on the continental margin channel off the west coast of Africa. Fuerteventura and Lanzarote, while still maintaining island status, are for all intents and purposes guyots on the way down. Fuerteventura is 16.5 Ma.

The point of this exercise was to establish whether wave cutting or subaerial erosion produced a guyot. For these islands and for the Hawaiian Islands, subaerial erosion and mass wasting are the primary agents of destruction. Guyot morphology is present even before the feature subsides below sealevel. In the life cycle of a seamount, the subaerial lifespan from the time the surface is broken until it erodes back to the surface appears to be about 16.5 Ma for the Canary Islands. The reason for that is that the age of Fuerteventura is 16.5 Ma, and it has nearly reached sealevel in its erosion/subsidence stage of evolution.

8. Great Meteor Group

Great Meteor Guyot (Figure 35) is the head of another of the seamount chains. Lying in the northeast-central Atlantic Ocean basin about 700 km south of the Azores, this is the only seamount chain on this azimuth in this basin.

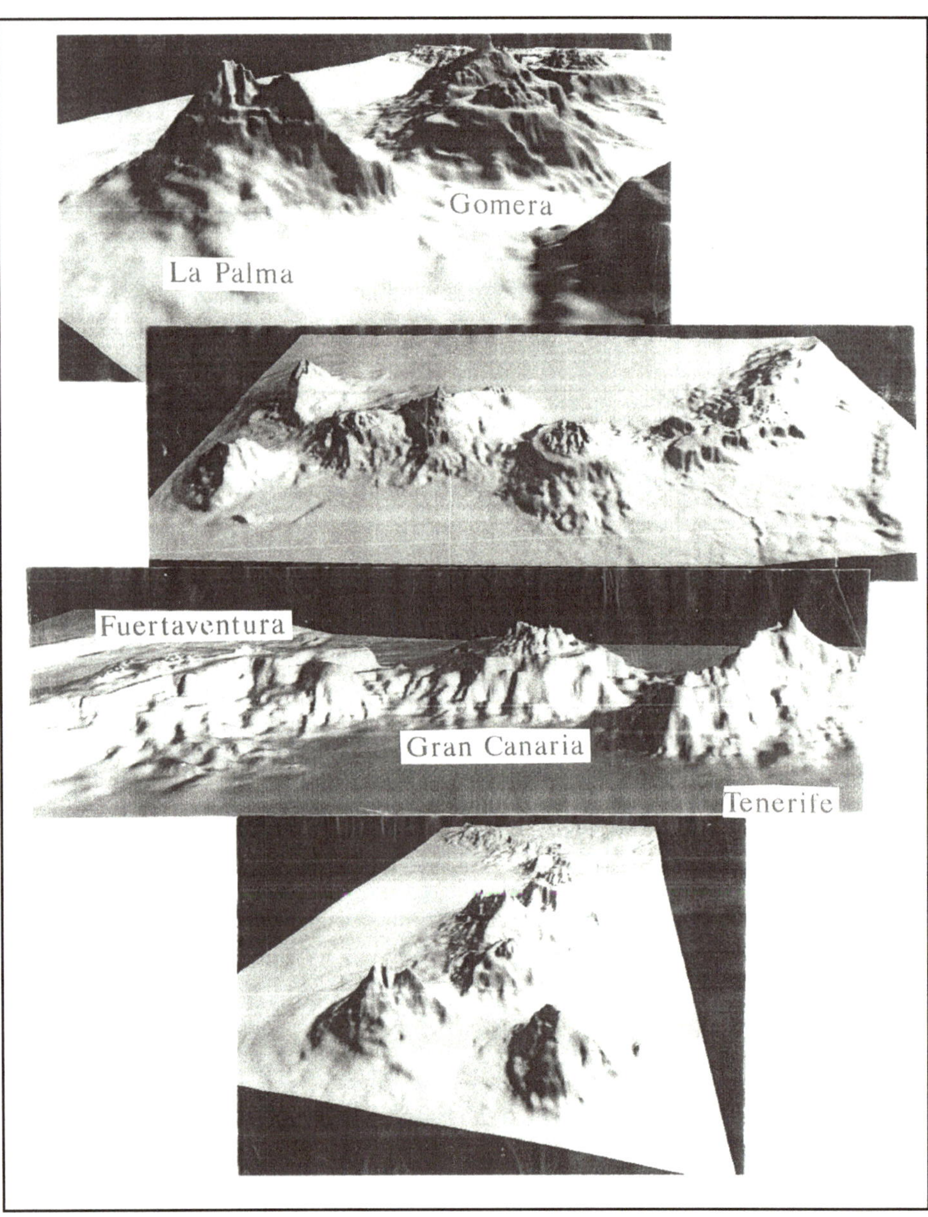

Figure 34. 3D of the Canary Islands in the Atlantic Ocean basin from multibeam sonar surveys of the bathymetry and topography from various maps. La Palma has a beautiful caldera. Gran Canaria is eroding to a flat top and Tenerife has a nice caldera.

It forms the southern part of the North Atlantic Gravity High. These seamounts are divided into four groups of NNW--SSE-trending lineations. After intersecting the eastern ridges of the Atlantis-Plato Seamounts of the Hayes Fracture Zone, the seamounts grow larger with distance to the south. This is a direct analog to the Emperor Seamounts. Cruiser Guyot is at 32°N latitude, 28°W longitude. Cruiser rises from the regional base depth of an average 2200 fm to a 200-fm summit on a 70-km long NNW--SSE trend. One would expect to find on a sediment draped feature, the notch in Cruiser Tablemount between Irving and Hyeres Seamount, and the similar platform depths of the two edifices now at 2500-2600 m. Subsequent volcanism has raised new depths to 1500-1700 m for the Cruiser Guyot and Piglet groups, 1000-1100 m for the Atlantis Seamount group, and 300-700 m for the Cruiser Tablemount and Plato Seamount groups. The name "Irving Seamount" is generally assigned to the feature on the NE portion of Cruiser Guyot, making this feature analogous to Broken-Top Guyot on the Michelson Ridge (Figure 23).

Irving rises to a 400-fm summit, and it's lineation is over 100-km long. Hyeres Seamount is at 31°30'N latitude, 29°W longitude. The summit of Hyeres is at 200-fm, and the lineation of 125° (NW--SE) runs for about 100 km.

Great Meteor Guyot finishes that lineation at 30°N latitude, 28°30'W longitude, 720 km south of the Azores and 1280 km west of Africa. Great Meteor rises from a 4800 m regional base depth to a 200-fm summit for a height of 4600 m. It lies on a NNW--SSE azimuth. This is the fourth lineation in this group of seamounts. Rocks recovered from the flanks of Great Meteor include boulders of schist and granite. The summit has yielded calcareous rock and biogenic calcareous sand. The high seismic P-velocity of 7.4 km/s and the high density of 3100 kg/m^3 of the core are not explained by basalts. They, on the other hand, fall more into the range of intrusive rocks such as gabbro and peridotite. The suggestion is forwarded that Great Meteor is indeed comprised of continental crust or upper mantle material.

The northwestern lower flank slopes on the Cruiser platform are excessively high. All of the other guyots in this book have steeper upper flank slopes that become more gentle towards the lower flank slopes.

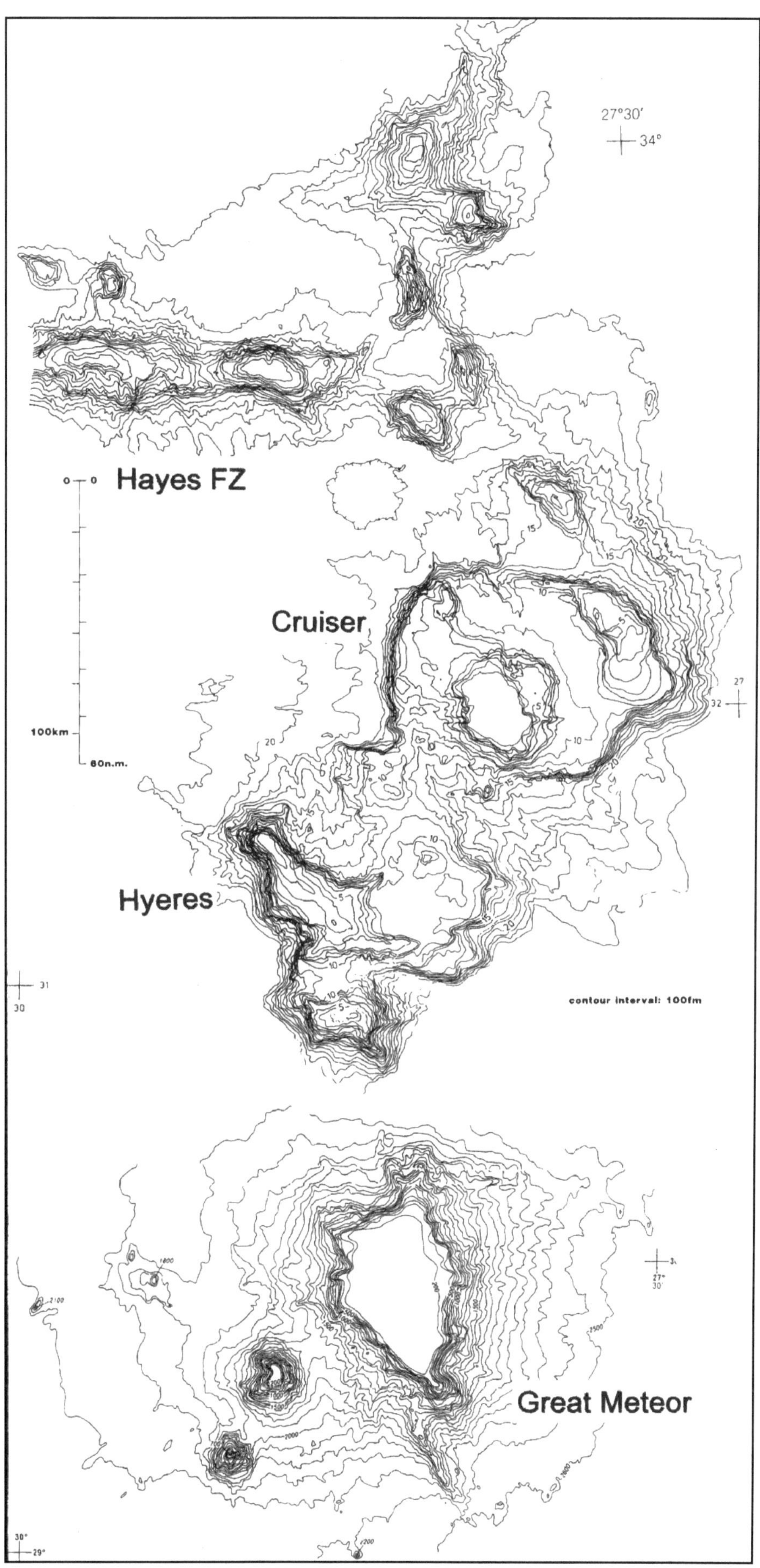

Figure 35. Bathymetry of the Great Meteor group at a 100-fm contour interval from multibeam sonar surveys. While the predominant lineation is N-S, each of the features has also been influenced by fracture control within the lineation. At the top, see also the Atlantis Seamounts on a direct E-W lineation. This is truly an enigmatic region and is not explainable by the plate-tectonic hypothesis.

The inclusion of Great Meteor Guyot into this setting introduces new and exciting possibilities. At once, the four N-S lineations through this megafeature show fracture control for the geomorphology. In theory, the ocean floor ages from 43 Ma at Atlantis Seamount on the north to 84 Ma at Great Meteor Guyot on the south. Only two reliable ages are extant for use here. Great Meteor is 11-16 Ma. Cruiser is early Miocene. The depths are apparently too shallow in this region to compare to any of the standard cooling curves for oceanic lithosphere.

MEGATRENDS

By now the astute reader has noticed some very interesting lineaments, or continuations, of certain features. Simply put, Figure 10 shows that most of the seamount chains in the North Atlantic are associated with the ends of the fracture zones. By comparing that figure to the GEOSAT diagram (Figure 11), any doubt is immediately removed. Where geochronology has been established, it is now possible to introduce a relatively new term, megatrends. Megatrends exist globally. Using satellite altimetry, coupled with bathymetry, these trends go all the way across all of the ocean basins, they lie on at least multiple azimuths in the Atlantic, Indian, and Pacific basins, and they totally refute any concept of seafloor spreading/earth expansion. Megatrends are underlain by hot lines, and they are characterized by both earthquake and volcanic activity. They cross trenches, ridges, entire basins, and even continents. Also, they are old. They pass through rocks that are Phanerozoic-Paleozoic in age.

And, the megatrends intersect each other on basin-wide patterns, at least four different lineations being shown for the Atlantic Ocean basin. For the Bullard fit to work, Massachusetts, USA, must fit with Morocco, Africa, and fracture zone(s) must join the two. The Hayes/Oceanographer fracture zone swarm, which includes the attendant seamounts, spans the region from Massachusetts USA through the New England Seamounts, across the North Atlantic basin, and on to the Iberian peninsula. On the western end of the fracture swarm the New England seamounts had been noticed to go ashore through Vermont to the Monteregian Hills in Quebec. The New England Seamounts/ Monteregian Hills were also known not to be age sequential, ranging in an admixture from 70 to 230 Ma. They lie on substrate that is at least 1 Ga. On the eastern extreme, the trend continued ashore as the Betic Cordillera crossing southern Spain. The basic meta-sedimentary rock there is Permian-Triassic in age. The fault zone associated with southern Spain is thought to be of a vertical nature rather than strike-slip. The megatrend is also in line with the Balearic Islands in the Mediterranean Sea.

Megatrends intersect, so it is now known that fracture zones do not point in the direction of seafloor spreading: the same piece of real estate can hardly be moving in two different directions at the same time. The plate tectonic explanation for the Pacific Ocean basin involves a myriad of twists and turns. A re-contouring effort of that basin using all of the available soundings to update that bathymetry (Figure 15) plus the GEOSAT structural diagram (Figure 16) gives meaning to the term “megatrend.”

Beginning in the northern basin, the classic definition fracture zones lie on a WSW-ENE lineation. Several are included to make the point. The Chinook Megatrend (Figure 15-3) has been shown to begin in Japan, cross the trench in the form of the Uyeda Ridge, continue through the Nadeshda Basin, underlie the Shatskiy Rise (Figure 15-4), pass through the Emperor Seamounts (Figure 15-5) and Hess Rise (Figure 15-6), cross the Emperor Fracture Zone (Figure 15-7) to become the Chinook Trough, and wind it’s way through the sediment traps of the Gulf of Alaska in the form of the three splayed seamount chain/ridges. One could draw the analogy that this has the geomorphic appearance of an eastward-flowing stream, what with the splaying on the eastern extreme.

Continuing south, the Mendocino Fracture Zone (Fracture Zone Chapter) now becomes the Mendocino Megatrend.

It would seem as though the eastward splaying is a rule for the megatrends on this lineation, which would necessarily indicate some sort of fluid mechanics at work in the tectonics rather than the movement of a lithosphere, “cast in stone” as it were.

The figures show many lineaments existing on the NNW-SSE axis. Generally, these are thought to be seamount chains created by hotspot activity where the Pacific "plate" has moved northerly before 43 Ma, turning abruptly, and almost instantaneously, WNW. Should these trends continue on the same axis, they will eventually have to cross the WSW-ENE megatrends.

Beginning on the SSE, the Udintsev Fracture Zone (Figure 15-18 and 23) actually continues northerly of the proposed Fiji plate (Figure 15-19), passes through the Dutton Ridge (Figure 23) where the Manken/McCann Ridge lies, and ends in the Ogasawara Plateau/Michelson Ridge (Figure 23) at the Izu Trench. This megatrend appears to have been "captured" in the center by the Fiji region, a region which has been hypothesized to be moving eastward by many researchers in a phenomenon called "trench migration."

The Kashima Megatrend (Figure36) lies to the east of that. Beginning on the north in the Geisha Guyots (Figure 15-20) at Kashima Seamount, this megatrend splays into two. One passes through the Michelson Ridge, the other just to the east. This double feature continues

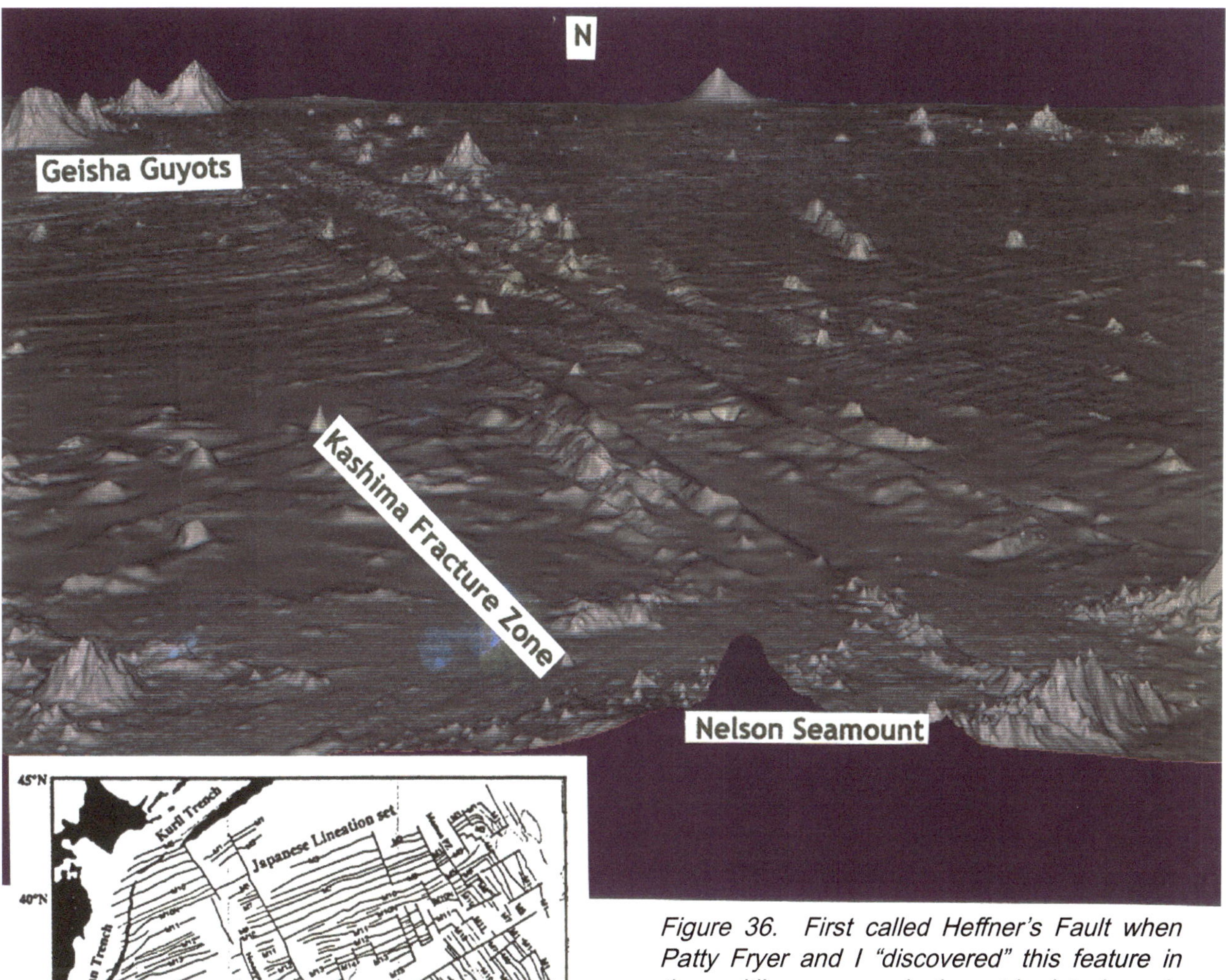

Figure 36. First called Heffner's Fault when Patty Fryer and I "discovered" this feature in the multibeam sonar bathymetric data base in 1986, the Kashima Megatrend 3D, lying in the NW Pacific basin, passes from the NNW to the SSE and clearly displays what is called tectonic spreading fabric perpendicular to the fracture zone ridges. That same morphology would occur if the region were undergoing compression due to cooling. This feature plays a huge part in later discussion, along with the Mamua and Krusenstern fractures in Figure 29.

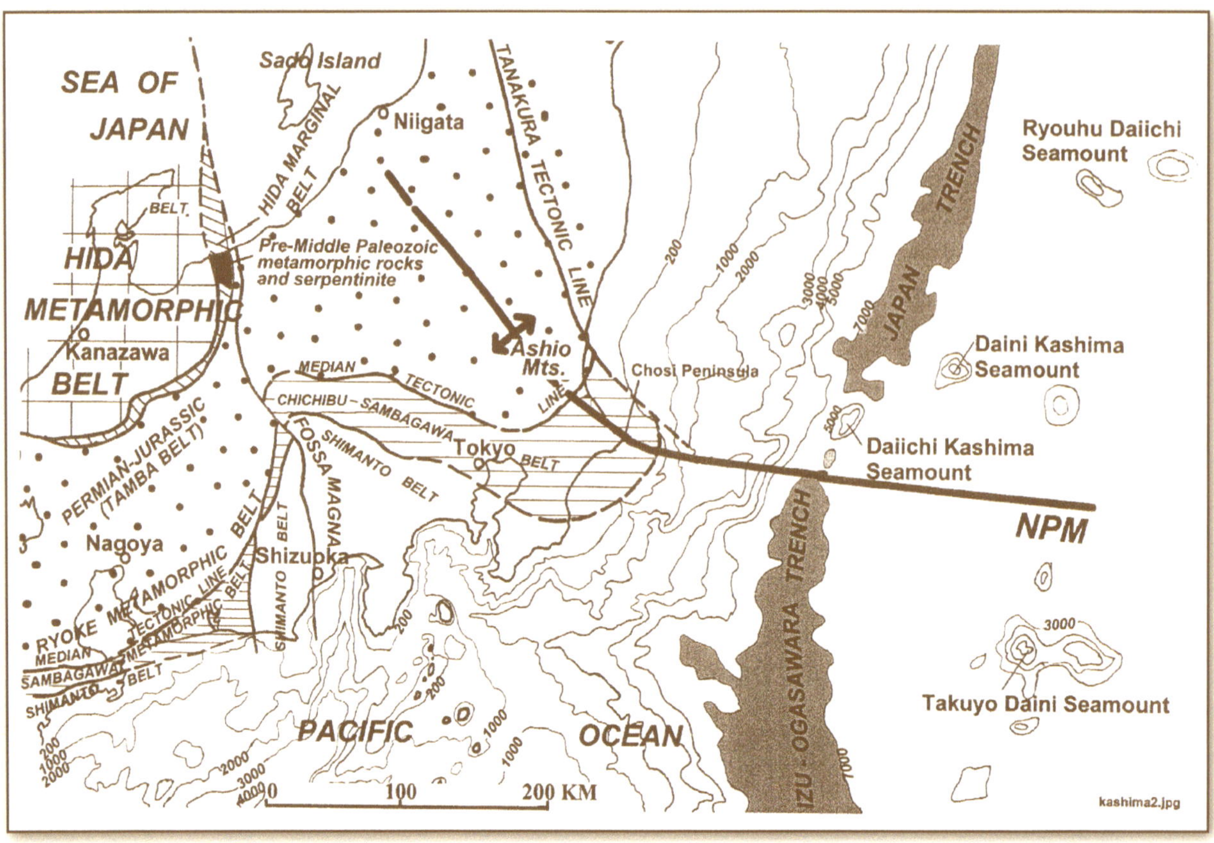

Figure 37. For the purposes of this discussion, the NPM begins in the Fossa Magna region of Japan, an area of Paleozoic rock ages.

southerly as the Marshall-Gilbert (Ralik-Ratik; Figure 15-21) seamount chains, is overprinted by the eastern Fiji region, passes into the Lau-Havre Ridge, and reunites in the form of the Louisville Ridge (Figure 15-22) and Eltanin Fracture Zone (Figure 15-23).

The Mamua Megatrend, displaying all the accouterments of a fracture with the perpendicular fabric, crosses the Chinook Megatrend and passes through the Mid-Pacific Mountains (Figure 15-24) as a discrete ridge. From there it is overprinted by the Manihiki Plateau (Figure 15-25), and it loses its integrity after going through the Tubuai Ridge (Figure 15-26).

The last Pacific basin megatrend to be discussed on this azimuth is the Emperor Megatrend. Beginning with the Emperor Fracture Zone (Figure 15-7) on the north, It continues SSE until it merges with the Krusenstern Megatrend at the Liliuokalani Ridge (Figure 15-28 and 19). A possible offset occurs just to the east as the ridge defining the western extreme of the Musicians Seamounts, which fades into obscurity both geophysically and bathymetrically at the juncture with the Hawaiian Chain.

These megatrends do not appear to be splaying at the eastern distal ends, which may be indicative of the age of the features comprising the megatrend.

Several WNW-ESE trending megatrends became apparent on the GEOSAT diagram for the Pacific basin (Figure 16). The first is called the North Pacific Megatrend (NPM), and the second is called the Central Pacific Megatrend(CPM).

The North Pacific Megatrend begins from Central Japan (Figure 37) where the megatrend is characterized by a conspicuous NW-SE geanticlinal high consisting of Paleozoic rocks. From there the trend slightly changes direction to WNW-ESE, and a continuous ridge is traceable toward the deep sea, south of Daiichi Kashima Seamount. The ridge separates the deep sea trench into two segments; Japan Trench in the north, and Izu-Ogasawara Trench in the south.

Crossing the trench reaches the Geisha Guyots (Figures 15-20 and 21). Daiichi Kashima Seamount at 35°50'N latitude and 142°40'E longitude, has been proven to be a tectonically undisturbed guyot with Middle Cretaceous (Aptian-Cenomanian) reef at the top. The seamount chain, although generally getting younger to the east, is not sequentially aged.

The Geisha Guyots have historically not been linked to anything else in the North Pacific

Ocean basin, possibly because of the gaps between the features. These gaps also occur in the more well-known Hawaiian Ridge and the Emperor Seamounts, so this is not a problem. However, they did not just mysteriously appear on a 120°/300° azimuth some 100 Ma, only to remain anonymous. On closer inspection, they possibly join the Hawaiian chain through the Mapmaker Seamounts.

After a short, buried segment, the megatrend emerges again as the Hawaiian Ridge (Figures 15-29 and 26). The Hawaiian Ridge begins with Colahan Seamount, the site through which the Mendocino Fracture Zone/megatrend passes. The Hawaiian Islands are shield volcanoes and shield volcano clusters. They are quite a bit larger than the Geisha Guyots. In fact, the seamounts/guyots all along this chain continue to get larger until the island of Hawaii.

From the Hawaiian Ridge, the NPM crosses the Clipperton and Clarion fracture zones, where the scarce bathymetry does not allow us to follow the passage. The GEOSAT data shows a rather prominent acute strike crossing those fracture zones to enter into the realm of the East Pacific Rise. In light of the ground-truthing of all the rest of the GEOSAT features, we will assume that this is the NPM and continue with the description.

Before continuing to the east between the East Pacific Rise and Central/South America, we need to catch up with another megatrend which will blend into this on, the CPM. The massive CPM begins for the purposes of this discussion the Banda Sea surrounded by five crustal elements: the Sunda Shield (Carboniferous age), Indian oceanic crust (Cretaceous), the Australian craton (Paleozoic), Pacific oceanic crust (undetermined), and a transitional complex (Paleozoic).

The CPM begins as a single 3-4 km-tall, 100 kilometer-wide ridge continuing out of the Banda Sea through New Guinea (the Maoke Mountains) and the Ontong-Java Plateau (Figure 15-12),

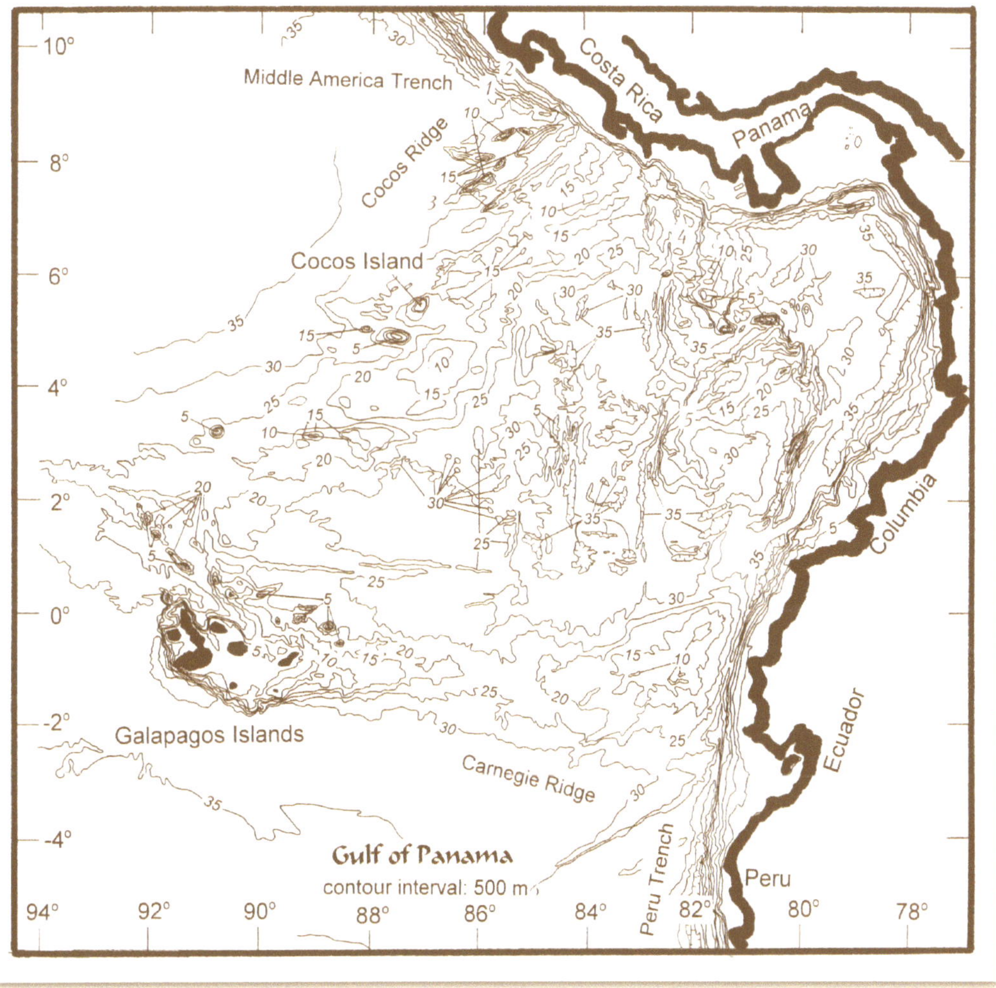

Figure 38. 2002 bathymetry of the Gulf of Panama/ Cocos "plate" at a 500-m contour interval and Mercator projection, compiled from ship-of-opportunity NGDC data base with an insert around the Galapagos islands from Bill Chadwick's 1994 chart. This is the region where the megatrend is joined by those from the north and south East Pacific Rise to accentuate the effects of the magma coming in from the WNW. The Cocos and Carnegie Ridges are the primary bifurcation. With compression acting on this feature, the (4) Coiba and (5) Malpelo Ridges form. Although they do not appear at this contour interval, (1) the Fisher Ridge and the (2) Fisher Seamount have anomalous ages. (3) Quepos Plateau has also been dated.

across the northern Fiji Plateau (Figure 15-19), and through the Samoan Islands for a distance of over 4000 km. This megatrend continues as the Tuamotu Ridge (Figure 15-31), the Easter Island Fracture Zone, and into the East Pacific Rise, where it splays into three segments (see cover for earthquake outline of this event).

The northernmost fork joins the flow at the East Pacific Rise to go northerly and join the NPM to become the Galapagos Ridge (Figure 38), which itself bifurcates into the Cocos Ridge on the north and the Carnegie Ridge on the south. The Cocos Ridge obliterates whatever trench may have been there and abuts Costa Rica. Then the ridge passes through Costa Rica to emerge into the Caribbean Sea as the Hess Fracture Zone/Ridge supporting the various islands starting with San Andreas and continuing through Jamaica and Haiti, a ridge between the Caribbean Sea proper on then south and the Cayman Trough (Figure 3-18) on the north. The southern segment includes the Galapagos Ridge and Carnegie Ridge (Figure 3-8). These features are supposed to be the divergent boundary between the Cocos and Nazca plates. The islands, also hypothesized to be a hotspot track, show a mixture of ages.

Several more ridges appear in the bathymetry. The Coiba Ridge trends more northerly, rises about 3000 m, and is a relatively short 200-km in length. The Malpelo Ridge, also arising from nowhere, passes northeasterly for about 400-km before fading into oblivion. It is broader but of lesser extent, rising only 1500 m.

Somewhere in this region in the older bathymetry was supposed to be a plate boundary, with the foregone descriptors applying to the Cocos "plate." I can see no reason to suspect such an event based on the updated bathymetry.

Additionally, from the Cocos Ridge to the south of the Carnegie Ridge, no trench exists in the bathymetry. The regional base depth appears to be about 3500 m, and several 4000 m readings occupy 2°30'N, 84°30'W and 6-7°N, 76-77°W. Subduction has been theorized for this region for the past 75 Ma. The earthquake regime is primarily above 200 km in depth, with no earthquake activity below 250 km. This means that no subduction can possibly be taking place from the Cocos Ridge on the north through 5°S, or from the Tehuantepec Ridge at 15°N to 5°S, at least for time covered by the data collection process.

Finishing the CPM, the central fork continues as the Sala y Gomez Ridge/Easter Fracture Zone (Figure 15-32). A study has carried that lineament onto the South American continent through the Nazca Ridge and onto the Precambrian Brazilian Shield, even though the sketchy bathymetry shows a subducting feature. The third, and southernmost, fork of the CPM starts on the south of the Ontong-Java Plateau, follows the same path, but splits off at Samoa to the south. It is comprised of the Austral Seamounts (Figure 15-33) coming out of the Fiji Plateau with Machias Seamount, the southern portion of the Society Islands, and it continues through south fork of the Easter Fracture Zone, the Juan Fernandez structure on the East Pacific Rise, the Chile Rise (Figure 15-34), South America, and the Falkland Islands in the south Atlantic Ocean. This megatrend passes through Precambrian rocks.

These megatrends all seem to splay on the eastern extremes, and they all appear to be the only active megatrends in the Pacific basin. So, they are at once both extremely young and extremely virile.

From this section the working hypothesis includes intersecting fracture zones/ megatrends for the Atlantic and Pacific Ocean basins. We have been given many rock ages, so that we now are able to more accurately establish a geochronology. But first, we will look at what happens at the megatrend intersections as we continue with the bathymetry portion of the geomorphology equation:

PLATEAUS AND RISES

One thing about the megatrends, they all intersect and underlie the rises, plateaus, and micro-plates. Where they intersect is, by definition, a large region of highly weakened lithosphere. Assuming magma is being delivered in some fashion to these regions, one would necessarily expect to find large, igneous provinces. That is exactly what happens. The region around the intersection becomes semi-molten and the magma pool expands; it is in tension. A swirl-like pattern appears in the bathymetry/topography. Where the excess finds a path to the surface, it produces rise-type, positive gravity anomaly features such as seamounts, aseismic ridges, volcano arcs, midocean ridges, rises and plateaus.

According to the USBGN: Rises are different than plateaus, and normally have different genesis. Plateaus and guyots may occur on rises. Rises are much more extensive and encompassing in areas, and delineate an area anomalous to a surrounding sea floor, which is always deeper than the 'rise.' 'Swell' was a term that was used in the community a long time ago. A plateau may drop off gently on one or more sides and a rise is gentle on all slopes. If there is a bathymetric difference, it appears to be like buttes and mesas on land with the plateau being smaller in the ocean floor's case.

Oceanic plateaus comprise about 2% to 10% of the total ocean floor area. Plateaus can have an interesting history in that they are variously thought to have formed as (1) continental fragments, or micro-continents, (2) basalt accumulations formed at leaky transform faults, (3) extinct island arcs, (4) an integral part of the lithosphere at accreting plate margins, (5) ancient spreading centers, and not as a secondary, midplate event, or (6) magma floods at the intersection of two or more fracture zones.

The better known plateaus have been in the literature since the early 1970s, and their general morphology has been shown by various mapping techniques. The names and locations of the recognized larger plateaus are taken from those atlases. The Atlantic Ocean and Gulf of Mexico contain the Heezen and Iceland plateaus.

The Pacific Ocean contains the Ogasawara, Walls, Ontong-Java (also called the Kapingamarangi Rise), Campbell, New Zealand, and Fiji plateaus, and the paleo-Dutton Ridge. The Arctic Ocean contains the Yermak Plateau. The Indian Ocean has the Naturaliste, Broken-Ridge, and Mascarene plateaus.

On the other hand, a rise is a swell, albeit a rather large one. Rises appear to be thermal in origin. The entire Hawaiian Island chain sits on the Hawaiian Rise. Mono, Corner, Nicaraguan, Bermuda, and Rio Grande rises are in the Atlantic Ocean basin. Other rises include the Eauripik, Greenland- Iceland, Azores-Biscay, and Carlsberg Ridge. The Malta Rise is in the Mediterranean Sea. The Pacific basin is home to the Hokkaido, Io, An, Hess, Shatskiy, Obruchev, Manihiki, Magellan, Lord Howe, and Three Kings rises. Last, the Arctic region has the Murmansk, Morris Jesup, and the Davis Strait rises.

The Magellan Rise is thought to be an abandoned spreading center. The Hess, Manihiki, and Shatskiy rises were all thought to have been formed by ridge crest volcanism. The Ontong-Java Rise was suspected to be a composite of continental and oceanic crust. Suspect as they may, all of these features formed atop leaky fracture zone intersections.

Called large, igneous provinces, various plateaus have been extensively studied. They are known to be from a few kilometers to a few hundred kilometers across. Originally all aseismic midocean plateaus were called micro-continents regardless of the rock type. The Rockall Plateau is known to be continental crust. The Cape Verde Terrace was formed by volcanism with depth anomalies explained by reheating of the lithosphere. Plateaus also exhibit weak to no magnetization, and this suggests they were not formed as oceanic crust. That point agrees with the thesis here exactly because all plateaus and rises form at the surge channel intersections by magma floods.

The positive gravity anomaly regions are home to an interesting phenomenon. Where the rises and plateaus have been noticed to lie atop the orthogonal junctions from both the bathymetry and satellite altimetry, geochemical results also verify this concept. In a study of over 50 of the world's flood basalt regions by Meyerhoff and Bhat, the chemical contents are found to vary the entire gamut of basalts, and also the gamut of ages from the present back to the Archaean. This includes andesites, trachytes, trachyandesites, rhyolites, and ignimbrites. Because of the chemical composition, the term magma flood is used. Magma floods comprise at least 68% of all the Earth's surface rocks. The same mechanism, therefore, caused such a range of features as MORs, linear island and seamount chains, oceanic plateaus, island arcs, and continental interiors.

All rifts have a pod of 7.0-7.7 km/sec material beneath the rift axes. Interestingly, all island arcs, foldbelts, and MORs have the same range. Because these pods underlie the extent of the feature, they must define a channel. The

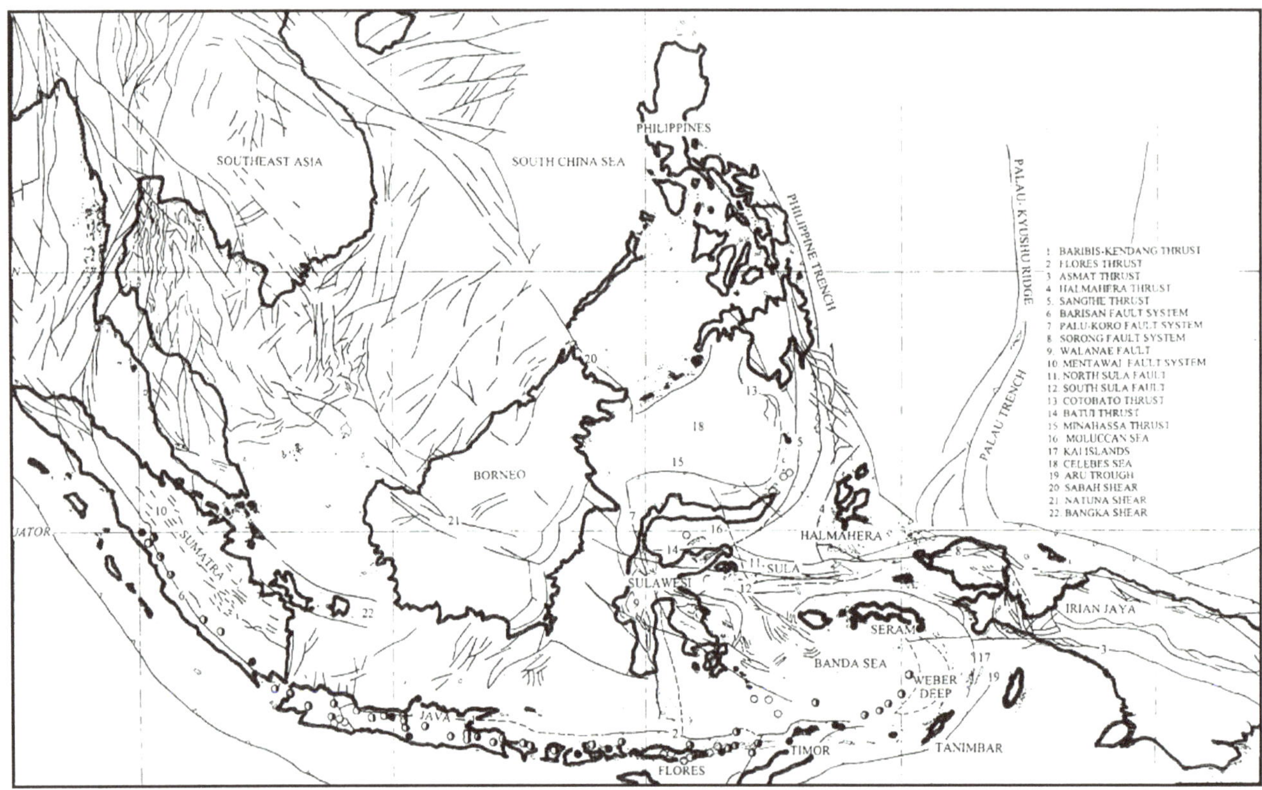

Figure 39. Compiled from a variety of sources related to SE Asian geology, this structural diagram shows the active faults and volcanoes in that region. Following the flow line, one can easily "see" the eastward trend out of the circular Banda Basin. This actually passes through New Guinea into the Ontong-Java Plateau.

channel is a low-velocity zone, so it must be in at least partial melt. When these channels intersect each other at an appreciable angle, the magma collides. In the region of the positive gravity anomalies, this collision produces a rupture through the overlying fault-fissure-fracture system which lets the magma rise out of the channel onto the surface. That eruption is the cause of the plateaus and other magma flood-type features. The hotspot and other explanations such as events at the MORs has now become merely a hot line, as it were.

1. Ontong-Java Plateau

The structural diagram for the Indonesian region (Figure 39) clearly shows the passage of a structural feature through Irian Jaya/New Guinea. This leads to the Ontong-Java Plateau (Figure 40) in the west-central Pacific basin. The southern portion of the OJP is also north of the western Pacific trenches, listed as the Vityaz Trench System (VTS) which extends over 2500 km in the form of the Manus, Kilinailau, North Solomon, Ulawan, and Cape Johnson trenches.

Presently straddling the equator, the OJP has been predicted to have formed by a South Pacific superplume between 90--150 Ma, near a ridge, and to have moved WNW to its present position from 43° S latitude and 160° W longitude. The OJP is embraced by the Udintsev and Kashima Fracture Zones. The feature which comes from the WNW and is in the center of the OJP. Another passes through from the ENE. This is the extension of the Nova-Canton Trough.

2. Mid-Pacific Mountains

The gross morphology of both the eastern and western sections of the Mid-Pacific Mountains (Figure 41) is of two rhomboids. This particular landform was noticed very early in the science of bathymetry collection and interpretation, and it was thought to be a compressional feature. Jerry Winterer of SIO and also the Hawaii Institute of Geophysics (HIG) were among the early investigators of the Mid-Pacs. In the updated bathymetry the rhomboid morphology changes to a whorl pattern. The close-order bathymetry of the discrete features comprising the Mid-Pacific Mountains is used to establish that fact. The features selected for this study are actually

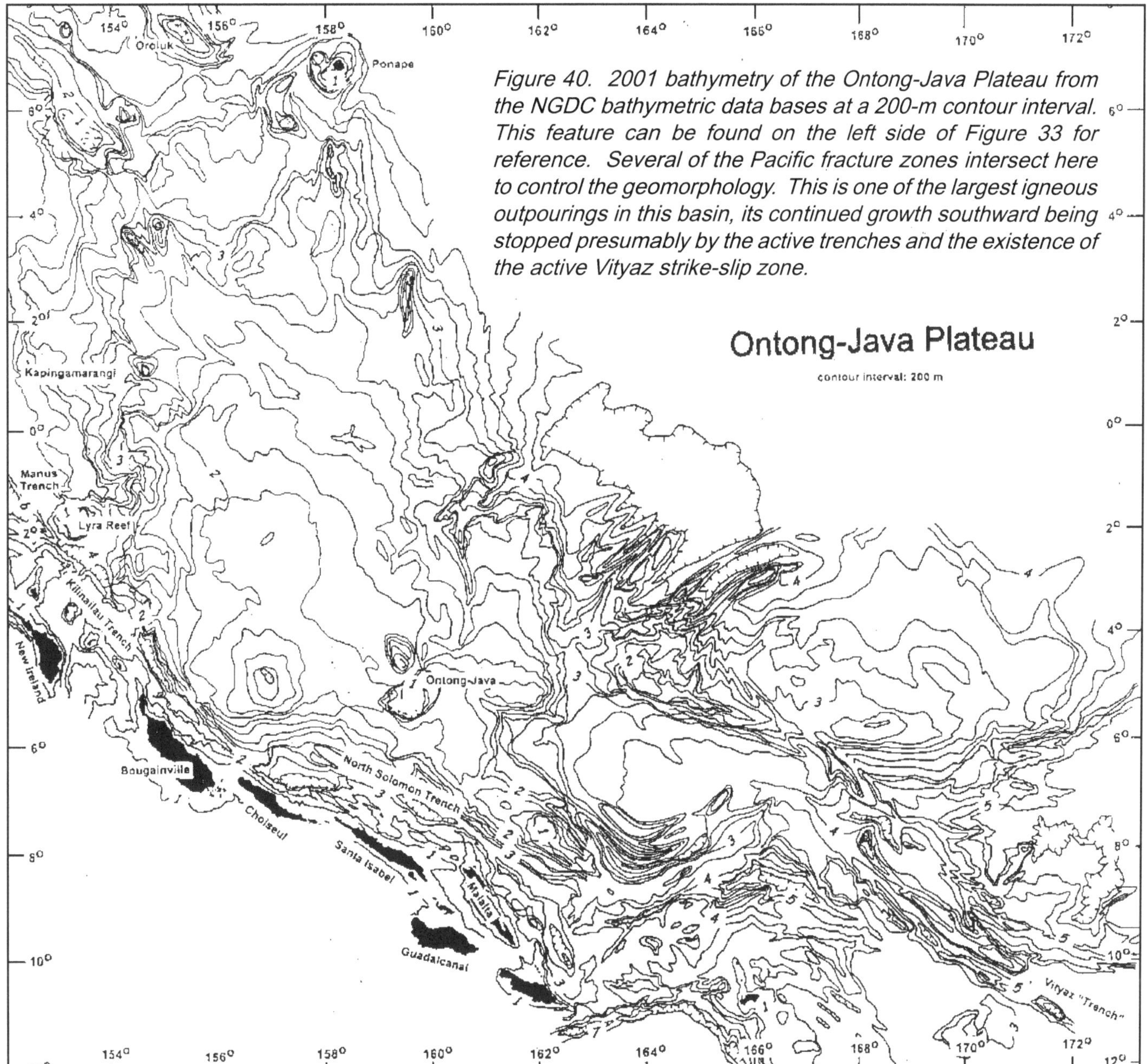

Figure 40. 2001 bathymetry of the Ontong-Java Plateau from the NGDC bathymetric data bases at a 200-m contour interval. This feature can be found on the left side of Figure 33 for reference. Several of the Pacific fracture zones intersect here to control the geomorphology. This is one of the largest igneous outpourings in this basin, its continued growth southward being stopped presumably by the active trenches and the existence of the active Vityaz strike-slip zone.

groups of features on volcanic pedestals that are associated with some common phenomenon, such as similar heights, alignments, etc. For the purposes of this study we begin with the younger, eastern section and work towards the older, western Mid-Pacs.

The entry into the Mid-Pacific Mountains is by the WSW--ENE-trending Necker Ridge and Horizon Guyot, which are the braided, anastomosing ridges of the Molokai Fracture Zone (Figure 42). Between 170°W and 171°W longitude this fracture zone is intercepted by the Line Islands, features which formed in and overprinted the NNW--SSE-trending Easter-Emperor megatrend transecting the entire Pacific Ocean basin. The WSW--ENE trend continues on a 235° azimuth to 178°W longitude, being overprinted by a pedestal at that point. Bathymetry associated with that trend demonstrates the underlying premise.

Fracture zone traces are for the most part fracture swarms, and that is the case here. Where seamount features are formed in pre-existing fracture zones, they take on the same alignment. Renard Guyot is on the 235° azimuth so prevalent for this region of the Mid-Pacific Mountains. A guyot/ridge to the north of Renard by 40-km and having the same characteristics as the Necker Ridge meanders through the Mid-Pacific Mountains bearing generally 255°. This splay in the fracture swarm leads directly to the next figure while reiterating the existence

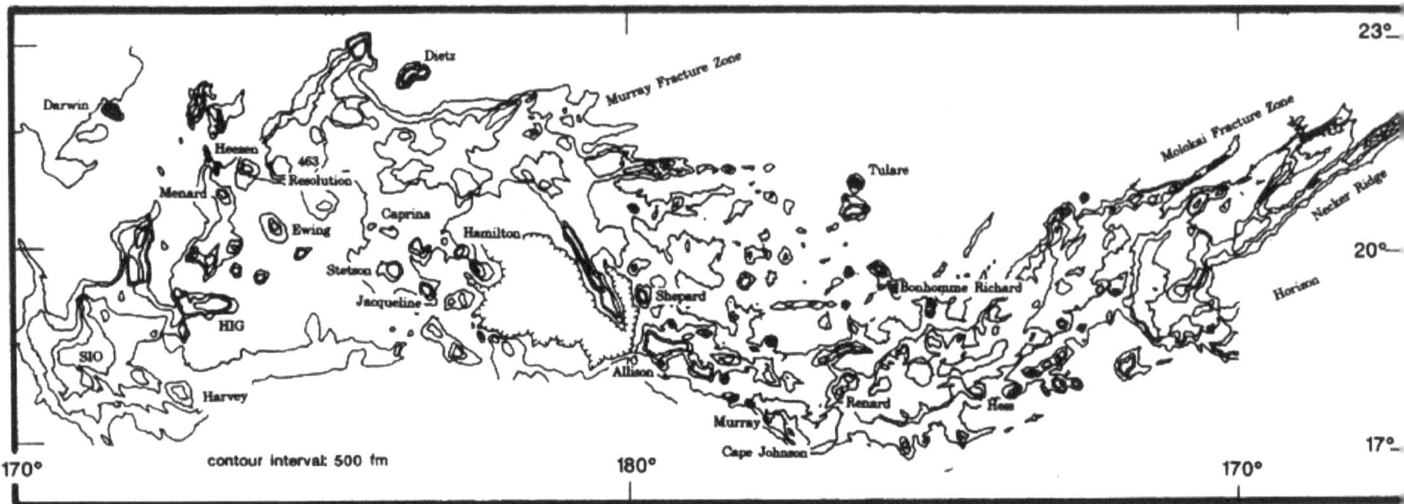

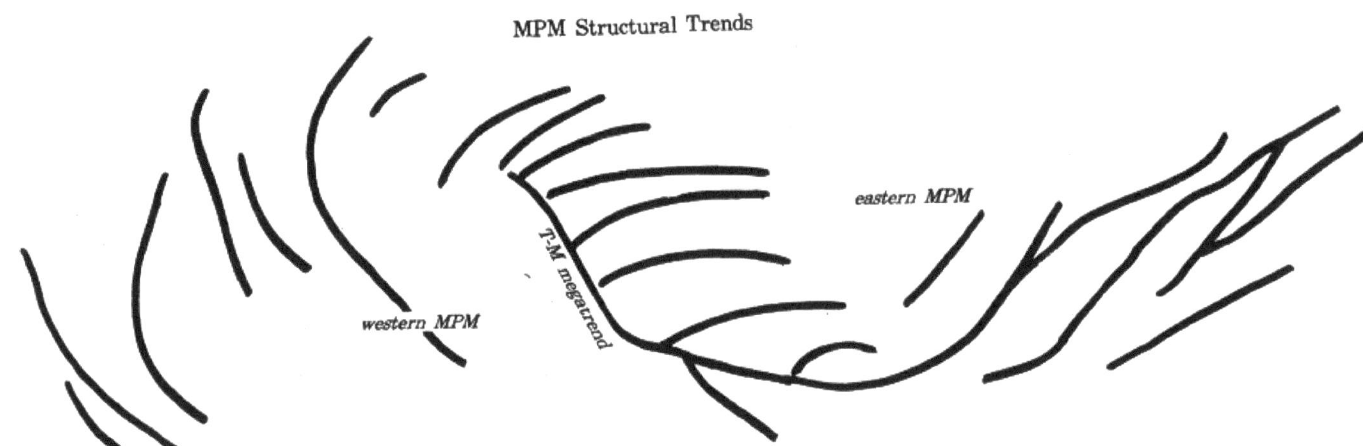

Figure 41. 1996 bathymetry of the Mid-Pacific Mountains at a 500-fm contour interval compiled from NAVOCEANO data bases. Outlining two semi-circular halves, the geomorphology of this feature was found to be fracture controlled by the Molokai and Murray fracture zones on the east intersected by the Emperor and Tubai-Mamua Megatrends.

of internal fracture control in the Mid-Pacific Mountains. The geomorphology of both of these ridges reflects fracture control on formation.

The next group to be shown in the eastern Mid-Pacific Mountains is headed by Allison Guyot. The diagram has two primary trends that must be addressed because this is the site of another orthogonal fracture zone intersection. This is where the Molokai Fracture Zone, bearing 235°, and an unnamed fracture ridge bearing 335° unite to distort the geomorphology of Allison Guyot. The 335° bearing fracture zone ridge is part of the Tubuai-Mamua megatrend. The orthogonal fracture zone intersection was in existence by 110 Ma because Allison Guyot has been dated at that time. The same 335° azimuth is on the NW portion of the unnamed guyot and carries north through the small seamount. Hess Guyot lies to the south of that.

The northerly trending Tubuai-Mamua fracture zone dominates the bathymetry of the central Mid-Pacific Mountains, continuing on an azimuth of 330° for 275-km. The Tubuai-Mamua fracture zone intersects the Molokai fracture swarm on the south and the Murray fracture swarm on the north at 21°N latitude.

In the northern region a rather interesting phenomenon occurs. The mini-ridges on the north of this figure begin to rotate counterclockwise. The ridge crossing 21°N latitude aligns at 300°, while the ridge at the top of the figure aligns at 290°. This rotation is a rule for the western Mid-Pacific Mountains, will be demonstrated in the rest of the 100-fm contour interval figures, and will be discussed later.

The incoming Murray Fracture Zone ridge at 22°N latitude and 179°E longitude trends 245° generally in a sinuous fashion for about 250 km.

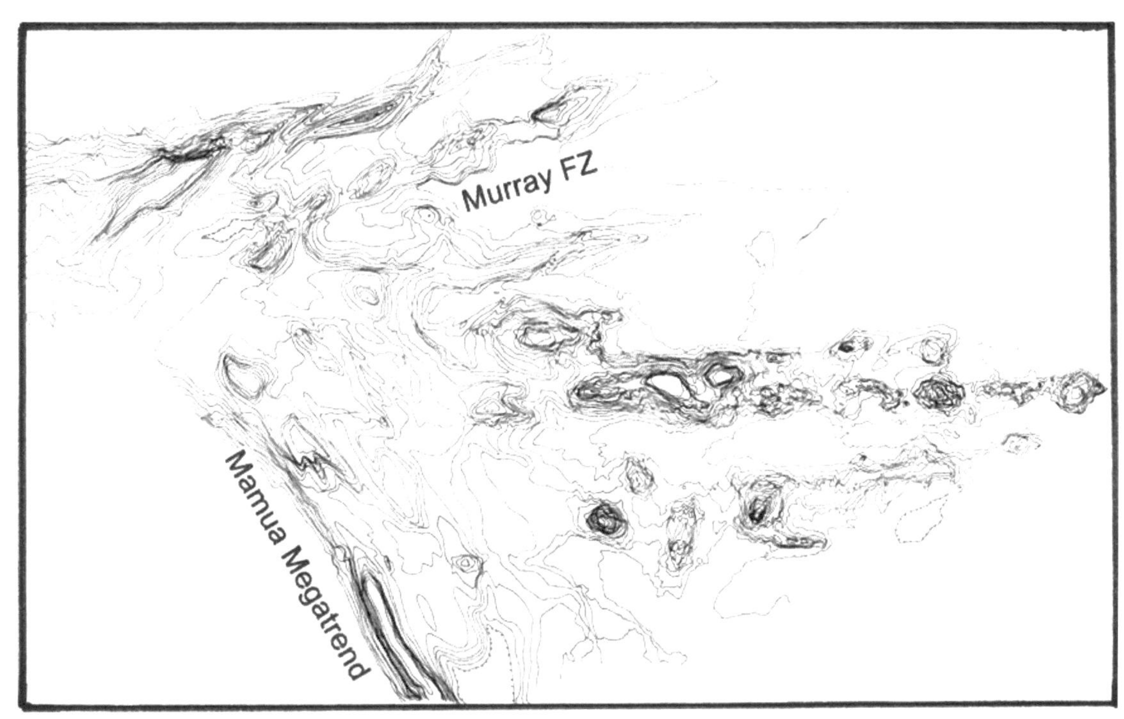

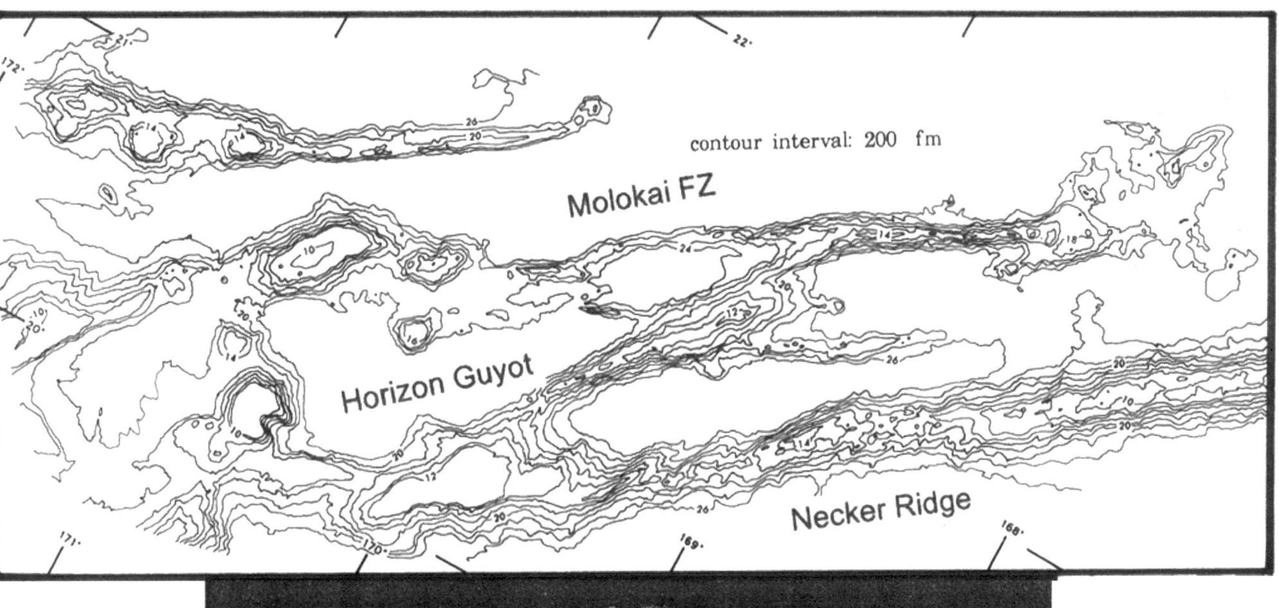

Figure 42. The fracture entering the Mid-Pacific Mountains from the ENE, this 3D compiled from NAVOCEANO data bases, shows the Necker Ridge/Molokai Fracture Zone as it intersects another from the NNW to have the eastern Mid-Pacs built as a plateau atop that intersection.

Figure 43. 1995 bathymetry of the Hess Rise from ship-of-opportunity data combined with earlier NAVOCEANO multibeam sonar survey data at a 200-fm contour interval. With four distinct sets of trends, this feature is fracture controlled. The Chinook and Mendocino megatrends intersect the Krusenstern and Emperor megatrends, while An unknown set of NNE-SSW trending ridges delineate the northern portion of the rise. Those ridges were discovered by NAVOCEANO in the 1980s, published, and have never been addressed again to my knowledge. They were surveyed with a total-coverage package.

It is also overprinted by a SW-trending pedestal, so that a determination as to its provenance is virtually impossible. In fact, the massive amount of overprinting in this region (Hess Rise, Shatskiy Rise, Mid-Pacific Mountains) makes a definitive statement about provenance almost impossible with the possible exception of the orthogonal fracture zone intersections. The two seamounts to the west of that ridge appear to outline a vortex. The guyot heights are higher on this platform, the slopes are gentler, and the summits are shoaler and older than on the eastern Mid-Pacs.

The next pedestal begins its curvilinear trend on a 220° azimuth, curves southerly toward mid-pedestal, and then curves easterly. Several of the guyots fall on a central line. The unnamed guyot at 22°N latitude and 175°25'E longitude appears to have been formed and compressed. The overall geomorphology is of an eastward-pointing horseshoe.

At the western Mid-Pacific Mountains extreme, the SIO Guyot complex, one can make a case for a 200° azimuth ridge on the NE of this figure curving east to 137°. The primary trends to the south all curve through south to 135° at three different sites on this platform as witnessed by the flank rift zones on SIO Guyot. Should the bathymetry be extended to the south, it would curve even more easterly.

ODP Sites 865 and 866 on the western Mid-Pacific Mountains, plus DSDP Site 171 on the eastern Mid-Pacific Mountains, show literally no lateral lithosphere movement from 90--128-Ma. This coincides precisely with the surge-tectonic theory where periods of orogeny, such as the great Cretaceous outpouring, correspond to periods of lithosphere collapse.

3. Hess Rise

Two rises are shown in the North Pacific. The bathymetry is based on a mixture of multibeam and ship-of-opportunity data, which renders sufficient detail to be included in a study of rises. The Hess Rise (Figure 15-16 and 43) lies between the Hawaiian Ridge on the south, the Emperor Fracture Zone on the east, and the Emperor Ridge on the west. The feature covers at least four degrees of longitude and five of latitude. It has been intensively sampled. The rise lies in the Cretaceous Magnetic Quiet Zone, the sole reason for which would seem to be because of the profusion and confusion of the magnetic lineations there. The bathymetry of single beam sonar produced a more than adequate rendition at first. The addition of multibeam data and a comparison of some of the older data has provided the "new" look at Hess Rise. The bathymetric provinces remain the same; i.e., southern volcanic province, central high plateau, NE flank, and western steps, only the definition changes. Hess Rise is broken by normal faulting into horsts, grabens, and fault scarps, which was happening on formation where the local rate of rifting could not keep up with magma production. The original shape was first altered by successive phases of tectonic volcanism. Hess Rise has the appearance of an inverted "T." The southern volcanic province lies on the exact azimuth of, because it is part of, the Mendocino Fracture Zone. A leaky fracture is the reason for all the volcanism. The inverted leg of the "T," the central high plateau, is on the same azimuth as other fractures surrounding the rise, Stalemate and Emperor in the immediate vicinity, and in fact lies atop Krusenstern Fracture Zone. Apparently, the Hess Rise could have formed at the orthogonal juncture of two fractures. The notation is made of the two vortex structures on the extreme eastern and western flanks.

DSDP Site 464 on the northern Hess Rise gives an age of 98-108 Ma. DSDP Site 465 on the southern Hess Rise gives an age of 103+ Ma, making the trough at least that old.

Of interest are the four sub-parallel ridges to the north, all trending to the NNE. Lying on no other lineament in the north Pacific basin, these seamount/ridges are unexplained other than to note that the seamounts are over one nautical mile high and present an enigma to any tectonic hypothesis. After having been introduced into the literature in 1984, they have not appeared on any later edition maps of the Hess Rise. This, too, is an enigma.

Figure 44. 1995 bathymetry of the Shatskiy Rise from Will Sager of TAMU on the west and NAVOCEANO multibeam sonar data on the right at a 200-fm contour interval. This feature lies at the intersection of the Chinook megatrend and the incoming seamount swarm of the Mamua megatrend.

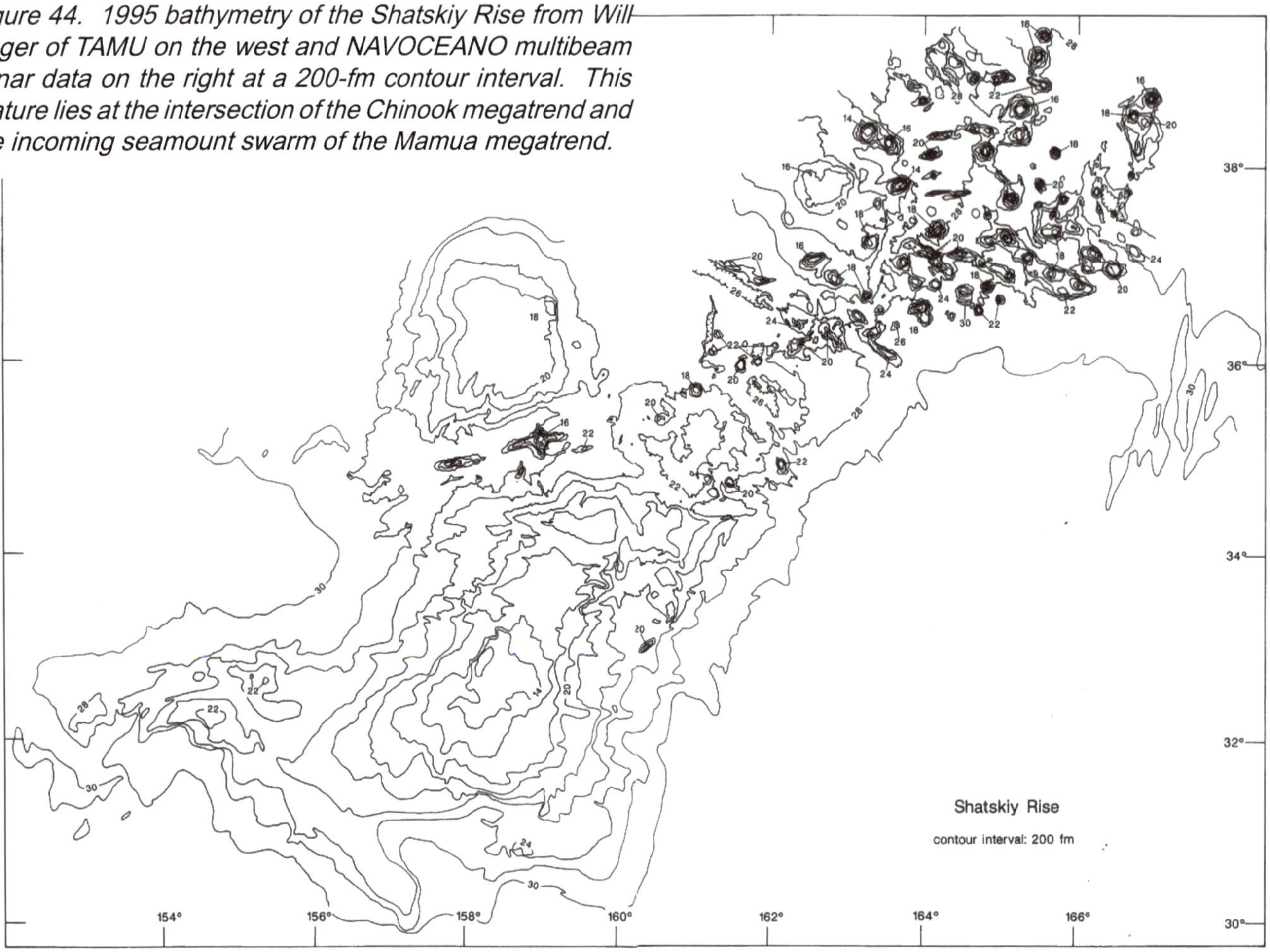

4. Shatskiy Rise

The Shatskiy Rise's (Figures 15-17 and 44) speculative formation has run the gamut; viz: an extinct triple junction, a microplate, large igneous province, and the result of a hotspot or a superplume. Will Sager of Texas A&M University (TAMU) led a team to map this feature, which data he combined with that of NAVOCEANO. The updated bathymetry shows a 1665-km WSW--ENE-trending feature to the west of the Emperor Ridge. The seamount province on the NE-quadrant trends NE, while the ridges made up of these seamounts trend NW. The rise appears to have a 220-km fracture ridge bisecting the central portion, so that it is divided into two segments. The Chinook Trough appears from the NE and reappears to the SW in the form of the Uyeda Ridge/Nelson Guyot complex, lending credence to this theory. Given the rise's overall morphology and its proximity to the leaky Chinook Trough fracture zone, the rise's formation atop and overprinting a pre-existing fracture is not entirely inconceivable. Which brings up a point about the rise's formation. Given the overall geomorphology and location along with the leaky fracture features shown elsewhere in the world's ocean floor basins, it is not entirely inconceivable that the Shatskiy Rise formed in, and overprinted, the preexisting Chinook Fracture Zone where it intersects several N-S-trending fractures, such as the Mamua Fracture Zone. The rise seems to be impervious to repeated drilling attempts, so that sequence of events can not be confirmed at this time. There are no basement-reaching holes on the rise.

Site 1213 (ODP; 2005) yielded a series of at least three basaltic sills with a total thickness of 46-m at the base of the sedimentary section. The units underlie limonitic breccias, evidence for hydrothermal activity associated with sill intrusion. The sills are interpreted as a widespread plutonic event associated with the latest stages in the construction of the Southern High of Shatskiy Rise. Sediment interbedded in the sills yields an earliest Berriasian age based on radiolarian and nannofossil assemblages and the sills themselves yield a mean ^{40}Ar-^{39}Ar incremental heating age of 144.6 “ 0.8 Ma. These data provide a minimum age to the origin of Shatskiy Rise, since the rise already had to be in place for sills to form.

Following several of the NNW-SSE trends produced an explanation for the existence and placement of all of the western Pacific rises. Udintsev Fracture Zone continued north through the recently-formed Lau-Harve region intersects with the Clarion Fracture Zone at the Ontong-Java Plateau. ODP Site 807 dates that feature at 120 Ma. Continuing north Udintsev Fracture Zone intersects Mendocino Fracture Zone, the site of Dutton Ridge. This region is undated. The next N-trending fracture off the “midocean” ridge is Eltanin Fracture Zone. As it changes into the Marshall-Gilbert Island trend, it intersects Clipperton Fracture Zone at 9°S latitude and 179°E longitude. That is the location of Solomon Islands Plateau, which remains undated. Further north the Marshall-Gilbert trend transitions to Heffner's Fault, and we have already seen that the intersection there is with Nelson Guyot. This is undated. The Tubuai-Manihiki Fracture Zone produces a veritable harvest of rises. The Tubuai-Manihiki Fracture Zone intersects the Galapagos Fracture Zone at 11°S latitude and 164°W longitude, the site of the Manihiki Plateau, dated at 125 Ma by DSDP Site 317. The next fracture intersection is with Clarion Fracture Zone at 7°20'N, 170°E. This is Magellan Rise, which has been dated by magnetics by Nakanishi and his crew (1992) at 120-130 Ma. The northern portion of the same trend is the intersection of Mamua Fracture Zone and Murray Fracture Zone. That region is home to the Mid-Pacific Mountains. They have been dated at 100-120 Ma, with the older portion on the western extreme by Jerry Winterer of SIO. The last intersection is with Chinook Trough, recently discussed herein as being overprinted by the as yet undated Shatskiy Rise.

Should the Udintsev Fracture Zone be continued to the north, it would intersect with Clarion Fracture Zone at Ontong-Java Plateau at 1°N and 157°E. The OJP is 120 Ma (ODP Site 807). The Udintsev Fracture Zone splays as it joins the Mendocino Fracture Zone at the Dutton Ridge. There are no dates for that feature. This exercise validates some of the geomorphology shown herein and disproves several of the theorized “microplates.”

ACTIVE MARGINS

While a study of active margins would encompass the seaward side, the trench, the landward side, the active arc, and the backarc basins in addition to the concepts of obduction and subduction, they are treated as a unit in this unit.

With every discussion the reader needs to master an esoteric **lingua franca**, and the topic of active margins is no different. Cdr. Balfour of the HMS PENGUIN took three soundings of over 5000 fm on the Kermadec Trench to record the earliest detection of a trench. These depths were noted in the HMS CHALLENGER report by Murray in 1895. The evolution of a active margin continued next with the discovery of movement along faults causing deep earthquakes in the vicinity of trenches by Wadati in 1935. The earthquakes were noticed to become progressively deeper with distance inland from the trench. Wadati's faults were interpreted as deep thrust faults where upper crustal blocks overrode lower blocks by Benioff in 1954.

The trench, then, is a function in and of the active margin. The active margin is comprised of the seaward lithosphere, the trench which has inner (landward) and outer (seaward) trench walls and an axial low, and the landward lithosphere. The outer trench wall is usually characterized by horst and graben structures from normal faulting and occasional aseismic bathymetric irregularities, such as seamounts. Inner trench walls can have terraces, perched basins, obducted fragments or ophiolites, re-entrants, and accretionary sediment prisms. The landward lithosphere contains the forearc and is overprinted by the active volcano arc in many cases. The width of this particular zone varies.

This region of the upper mantle is known variously as the subduction zone, the Benioff zone, the convergence zone, or the active margin. The subducting slab is another of the driving forces of plate tectonics, called “slab pull.” Continental and oceanic volcanic arcs form landward to Benioff zones and are the foci of extremely virulent volcano and earthquake activity. Through time, the ridges, transforms, and trenches may move about in no organized manner producing such phenomena as plate reorganizations, changes in plate movement direction, polar wandering, magnetic shifts, trench migration, and others. The third driving force of plate tectonics is the convection cell underlying the ocean basins. Your attention is directed toward the active margins:

By definition, several types of active margins exist. (1) In a collision between oceanic lithosphere, one subducts and the other overrides. The subducting plate is called "seaward," and the overriding plate is called "landward." (2) In a collision between a continental plate and an oceanic plate, the oceanic plate always subducts because the oceanic lithosphere is thinner. (3) In a collision between continental and oceanic plates where there is also oceanic crust present seaward of the continent, the landward oceanic crust can subduct the seaward oceanic crust first. Finally (4), two continental plates meet in a continental collision. This process gives a suture zone, or mobile belt, which is a closed subduction zone. A large amount of horizontal compression and over-thrusting always exists in these settings. All of the intervening oceanic crust is consumed, followed by the abandonment of the subduction zone. The shallow sea in between is filled with sediments, and the mountains begin to rise. These sediment deposits are folded, and igneous rocks are intruded into them.

The convergence process affects trench inner wall and forearc geomorphology. Parameters governing the amount of the affect include the convergence rate, the degree of sediment thickness on the seaward lithosphere, the degree of development of normal faults, horsts, grabens, and bathymetric irregularities on the ocean-side lithosphere, the type and intensity of trench axis and inner slope deformation, lower slope and trench sediment accumulation rates, and sediment thickness within the trench. The subduction process may be temporarily terminated by the introduction of major bathymetric highs. The collision of large, aseismic, bathymetric highs will permanently alter the trench's geomorphology.

The active volcanic arc lies landward of the trench. It is the second main feature of the active margin. Volcanoes are ubiquitous on the ocean floor and are well-known in the literature. Of Earth's volcanoes, 80% occur at active margins. The two primary sites on Earth are the Pacific "ring-of-fire" and a belt stretching from the Mediterranean through Asia to Indonesia. The volcanoes of Central America and Mexico, the Cascade Mountains of the western United States, the Aleutian Islands, the Kuril Islands and Japan, and the Philippine Sea arcs comprise the northern Pacific ring. South America has the Andes Mountains. New Zealand and Tonga are in the SW Pacific. The only active arc in the North Atlantic is the Antilles arc, including the islands of Martinique, Domenica, and Antigua. The Scotia arc lies off southern South America in the South Atlantic Ocean. Italy, Greece, and the surrounding islands are part of the Mediterranean arc. The Indian Ocean has the Indonesian arc with more than 70 volcanoes in its 3000-km length.

Most active arcs are comprised of nearly straight segments cut by transverse fractures. This fact was first noted by Ted Ranneft in a 1979 paper. He was an exploration geologist all over SE Asia during his lengthy career. Where there are strike changes in the arc there are multiple transverse fractures. This has been found to be true for almost all of the active arcs. The transverse fractures may continue to the trench in high stress situations, such as those brought on by the introduction of a large seamount into the region.

One rather prominent feature is associated with active margins, and that is the cusp. While only a few exist in nature, numbering probably less than fifteen worldwide, they are large enough to elicit comment. First noted by Peter Vogt and his co-workers in 1976 to be where large aseismic oceanic ridges were residing on the seaward lithosphere, they were suggested to be the effect of subducting these ridges, ridges which appear to be resisting subduction in the plate-tectonics scheme of things. Many of these cusps will be discussed. These include the two continental cusps surrounding the Indian sub-continent as well as those lying in the northwest Pacific Ocean basin. While the final disposition of the aseismic ridges in their collective dotages may be the same; i.e., they will disappear as an entity, the emplacement mechanism will change appreciably.

Having discussed the different features that comprise an active margin, this observation is

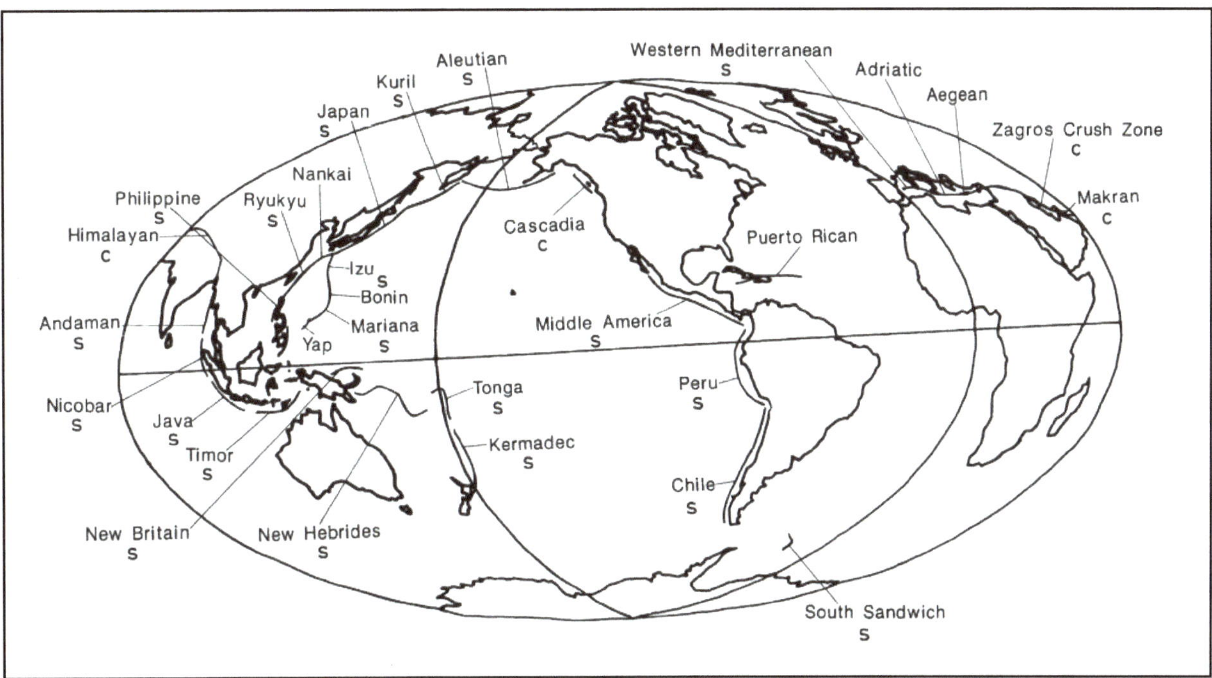

Figure 45. Convergent margins of the world where "s" is subduction zones and "c" is collision margins.

made: nearly all of the active margins are in the Pacific basin and face the east (Figure 45). The mechanism for this phenomenon was postulated to be an eastward flowing magma stream existed in the upper mantle that was connected to the convection cells beneath the midocean ridges.

The trenches are usually underlain by earthquake activity. The earthquakes appear to lie in at least two planes if intermediate or deep earthquakes exist at that point. All of the active margins have numerous earthquakes in the upper 100 km. Earthquake profiles across most of the trenches show what is happening at depth. These are shown at the end of the chapter. The rock types, while not having actually been samples at depth, are able to be extrapolated from multi-channel seismic data. These profiles are also shown at the end of the chapter in order to simplify matters.

The intent of this treatise is to collect all of the known information concerning active margins. Where possible, multibeam bathymetry is presented for the geomorphology. Earthquakes played a large part in the defining years, so they are used. Any available rock ages and seismic profiles were gathered. Because the information is being gathered so rapidly and may be subject to change at any time, two Appendices are attached at the end of the book. They are Appendix B: Earthquake Profiles and Appendix C: Seismic Profiles. Their descriptors are in the proper sections of the following chapter. The last of the primary tectonic features, they tell a story in themselves:

1. Kamchatka Cusp

The Aleutian and Kamchatka/Kuril trenches meet in the extreme NW Pacific basin, and this is the site of the Kamchatka cusp. The first feature, the Obruchev Rise (Figure 29), overprints the northern extreme of the Emperor Fracture Zone. The Rise is bounded by portions of NNW-SSE trending megatrends, the Stalemate Fracture Zone on the east and the Krusenstern Fracture Zone on the west. There is a small amount of left-lateral offset. Obruchev Rise presents a 130-km front to the trench. With 2560-m of relief with 3% slopes, this feature should not have much of an adverse affect on the forearc and the earthquake regime and, in fact, does not. The fracture zone ridges also have a negligible effect, at least from a geomorphologic standpoint. The Krusenstern ridges are 732--915-m high. The geologic impossibility of such an event notwithstanding, the 1830 m Stalemate fracture ridge appears to have caught a snag in the Aleutian Trench as it passes parallel to that trench and is being "peeled" to the east.

The Kamchatka forearc looks like a display case for canyons. The canyons follow fault traces. Subaerial erosion scoured out the canyons during the Miocene to later.

The Kamchatka Peninsula lies at the northwestern limit of the ring-of-fire. The Kamchatka volcanoes are in larger clusters than in the surrounding regions. At 250 km landward of the trench axis, the Kliuchevskaya group contains 12 volcanoes, most reaching heights over 4575-m (15,000 feet), including Kamen, Kliuchevshoi, Sheveluch, and Bezimianny. Where large, aseismic, buoyant bodies are in the trench, the rule is more deep earthquakes and more clustered volcanoes on the active arc. In keeping with the theory of activity, Kliuchevshoi has erupted 86 times since 1697 and is even now putting off gas and steam. Sheveluch had its largest eruption ever in 1854, one that destroyed a cone in an older caldera, formed another caldera, and caused a large debris slide. It now has a 10-km wide crater formed in 1964 housing a lava dome which expanded in 1980-81. The 1981 eruption was not preceded by seismic activity, presumably because it did emerge in that same crater. Phreatic explosions followed in 1984 and 1987, with the final one in 1991. Bezimianny is a 9500-ft stratovolcano that formed about 5000--5500 years ago over an older volcano. In October 1993 it had its largest eruption since 1956. Several other volcanoes in this sphere of influence include Zhupanovsky, a four-coned basaltic-andesitic complex. It has undergone seven eruptions since 1776, the most recent of which occurred in 1959. Avachinsky is now undergoing fumarolic activity, having started forming 60--70-Kyr. A large debris slide occurred 30--40-Kyr, and 640 years ago a powerful explosive eruption produced a large caldera. A massive basaltic shield volcano called Tolbachik is located at the south end of the Kliuchevskaya group. It is composed of two overlapping volcanoes with a lengthy rift zone extending to the northeast (Smithsonian Institution's Bulletin).

The rise and the two fractures have little or no effect on the earthquake regime, presumably because the seismic phases yield values that are NW-NNE, and their distribution is strongly at right angles to the convergence zone. The shallow Benioff zone dip angle is about 55°, which is constant from 50°N to 55°N. The earthquake regime does not show a grouping of the deeper ones, showing instead a gradual gradation over the entire Benioff Zone behind the trench (Table II; Appendix B-5). This is attributed to the location, an acute cusp, and the geomorphology of the Rise itself, low side slopes. Then, at 750 km behind the trench, there is an indication of a vertical Benioff zone shown by the scattering of deep earthquakes, a "red zone" [red being the color used herein to show earthquakes deeper than 300 km B-4 is reversed.]

Table II. EARTHQUAKE DATA FOR THE KURIL AND JAPAN TRENCHES.
(data source is the NGDC 1990 earthquake CD-ROM)

Earthquake depth (kms)	Quantity
0-49	9814
50-99	2807
100-149	705
150-199	246
200-249	122
250-299	123
300-349	77
350-399	65
400-449	60
500-549	46
550-599	32
600-649	23
650-699	6
700+	0
unknown	603

Dredge hauls made on Detroit Guyot, a feature residing on the Obruchev Rise, have yielded Precambrian ages. Boris Vasil'yev reported (1992) metamorphic rocks from what had been interpreted as basaltic oceanic crust. The continental rocks are overlain by Late Cretaceous acid-volcanic and sedimentary rocks, possibly glacial in origin, of Pliocene age onto which subaerial volcanic effusives flowed. Numerous dredgings were taken on the seaward slope of the trench and on the Zenkevitch Rise in water 8500--9000-m deep and water 5400 m deep respectively. The recovered rocks are crystalline garnet schist and pyroxene from the base of the trench. Freshly broken surfaces indicate an in situ origin for these samples. The crystalline schist has been originally thought of as oceanic layer III. The basement complex is now mostly covered by thick (2--3-km) Late Jurassic--

Early Cretaceous pillow lavas (tholeitic basalt) intruded by hyper abyssal dolerite, gabbro-dolerite, and gabbro-anorthosite of Late Cretaceous age. This led Vasil'yev to conclude that the Obruchev Rise had been analogous to the Kamchatka peninsula, having the same rock types. He stated affirmatively that the Obruchev Rise was continental in origin, and that it had undergone rapid submergence during the Pleistocene.

Seismic profiles taken across the Kuril Trench (Appendix C) show a stratigraphy in which the basement, Unit I, is folded and clearly layered, though wavy and discontinuous. It is block faulted, one of which is situated at the center of the trench and appears to be active during the Cenozoic. This is affecting the ocean floor bathymetry. Unit II thinly drapes Unit I. Units I and II are considered to be the source of dredged metamorphic, mafic, and plutonic rocks which are Precambrian in age. Unit III formed the bathymetric high in the present day trench area at the time Unit IV was deposited. Unit IV thickens conspicuously toward mid-slope up to almost 2 km. A large subsidence took place after the deposition of Unit V. This is presumed to be the normal fault which formed the trench.

Trench segmentation and reliable seismic traces across the trench reveal that no subduction is occurring here. They show, instead, a continuous substrate from the landward to the seaward side. The trench is segmented, so there is no continuum of processes acting here. We were able to get one good measurement here, that of 7000 m for the trench axis maximum depth.

2. Hokkaido Cusp

The next feature to the south at the active margin is Erimo Seamount (Figure 46), which has been surveyed using the SeaBeam sonar system by the Japanese Hydrographic Office. Erimo Seamount is in the trench axis at 41°N latitude, the juncture of the Kuril and Japan trenches. It does not show the usual signs of normal faulting for a seamount in this setting. The profiles show a simple pattern in that there is no over-thrusting or stress transfer to the inner trench wall. The lithosphere appears to be down-faulting as it bends, but with no sonar coverage there, that is conjecture. The trench axis is 7300-m deep.

The earthquakes are typical for a seamount in such proximity to a large compression feature such as a trench, showing a cluster of epicenters 250 km inland from the trench (Appendix B-5). The Benioff Zone dip angle is 33°. The earthquake foci in the vicinity of the Kuril and Japan trenches show that the shallow earthquakes are densest at the mid-continental slope, some 70--140-km landward of the trenches. This is not where the Pacific "plate" is presumed to be starting to subduct at the rate of 10--10.5-cm/yr.

gure 46. Erimo Seamount at the Kuril-pan trench cusp at a 100-fm contour interval rveyed by the Japanese Hydrographic ffice in the 1980s. While the lower flanks e in the trench, they show no signs of econdary and tertiary geomorphology, and e seamount lies relatively undisturbed at e trench axis. Not so for its neighbor to the uth at the juncture of the Japan and Izu enches, called Kashima Seamount. Even a 100-fm contour interval this small feature eems to have downdropped in the middle to the trench, having little-or-no deleterious fect on the surroundings.

The region has been well-sampled by dredging and seismic surveys. Vasil'yev and others (1986) dredged between 42°N--43°N latitude and 151°E to 152°E longitude at water depths from 8500--9000-m and at 5400-m. They found garnet-bearing slatey rocks and pyroxenite, which are continental or mantle rocks. These freshly broken surfaces had previously been thought of as oceanic Layer 3 by seismic surveys. DSDP Site 436 south of Erimo Seamount reveals radiolarian cherts at the bottom. Seismic stratigraphy (Appendix C) shows a 4-5 km thick basement layer that includes Archaean and Archaean-Proterozoic material. They conclude that the oceanic crust at the Kuril and Japan trenches is not oceanic, but terrigenous clastics with a westward progradation. Vasil'yev and Evlanov (1982) found continental rocks on Takuyo-Daiichi Seamount, Ryofu-daini Seamount (NE of Kasima Guyot), and Erimo Seamount. These were found in situ with basalts and considered to be the acoustic basement in this area. This led to the conclusion that the "oceanic" crust here actually consists of Precambrian and younger continental crust by Dong Choi, Boris Vasil'yev, and Ismail Bhat in 1987.

3. Izu-Bonin-Mariana Region

The Izu-Bonin-Mariana (IBM) active margin has been intensively studied. The IBM landward lithosphere is characterized by the lack of a sizeable clastic accretionary wedge, a mafic igneous nature of the forearc basement, a thin sediment cover, and the presence of sedimentary breccias on the inner trench wall. The stress regime with this combined front has caused the alteration of much of the primary geomorphology on the landward side.

SeaBeam surveys over Kashima Seamount (Figure 46) show a downfaulted guyot at 36°N latitude, the juncture of the Japan and Izu trenches. The 55-km diameter guyot is being reduced in a normal fashion with no exotic material being plastered on the inner trench wall. The surface of the guyot has been intensively studied, showing a large scarp caused by normal faulting associated with the subduction process. The vertical displacement of the lower flat surface is 1290 m. The surface was reef-capped in the Early Cretaceous. The western block is somewhat tilted, leading to the belief that a larger edifice has preceeded Kashima; that is, that the remaining part is just that, a part of a larger feature.

An ENE-trending fracture ridge at 34°N latitude is 130-km long ridge and, in spots, is up to 1100-m high. The ridge strikes the Izu Trench at an azimuth of 042°. The trench is barely affected by the introduction of this ridge. An inner trench wall terrace is perhaps formed by sediment deposition by the forearc canyon system shown on the figure. The earthquake regime gives the 33° dip angle which is so prevalent for this region, with some small amount of deeper earthquakes 520 km landward of the trench.

At 32°30'N latitude, the 42-km diameter Delaine Seamount shows an assortment of faults. The first downdropped portion of the seamount in the center of the Izu Trench is oriented 340°. This segment is normal faulted 730-m at a 36° angle from the next plateau upslope. The central fragment is oriented outwardly from the main edifice, ranging from 355° on the south to 016° on the north. This fragment has downdropped 915-m from the upper section. No accretion occurs on the inner trench wall. The earthquake regime shows a 33° dip angle on the upper portion of the Benioff Zone and a 75° dip angle on the lower. This is one of the rare instances where this phenomenon occurs so clearly. Since most of the deep earthquakes for this region occur between this point and the Ogasawara Plateau, this phenomenon is probably a function of the proximity to the Ogasawara Plateau. An earthquake cluster lies at 29°N latitude, with more between Kamchatka and Sakhalin. The deeper earthquakes are further inland from the active arc.

The active arc passes through Japan, a region of intense volcanism and earthquake activity, which is partially outlined by the Fossa Magna region. As the active arc passes back to the ocean basin, it is called the South Honshu and East Mariana Ridges.

The South Honshu Ridge lies 250 km from the Bonin Trench axis. The earthquake regime is primarily in the 250-400 km range from 27° to 32°N latitude under the entire South Honshu Ridge, with a host of shallow earthquakes centered at 28°N latitude and 140°45'E longitude. The volcanoes appear as clusters, specifically

Figure 47. Forearc, trench, and seaward Ogasawara Plateau from multibeam data at a 100-fm contour interval.

just to the north of the Volcano Islands at Kaitoku with at least three tops, Kaikata next with its cluster of five, and Nishino Jima, whose eruptive sequence in 1973 has been shown. The island has since eroded back to a submarine volcano. The andesitic Kaikata is dormant, and Kaitoku is active. Iwo Jima is comprised primarily of alkalic volcanism. The southern seamount province erupts mainly basalt through dacitic rocks. The central island province erupts primarily basalts and basaltic-andesites.

The cusp at the Ogasawara Plateau region is located between 25°N and 27°N latitude at the juncture of the Bonin and Mariana trenches about 1620-km south of Japan and 325-km east of Iwo Jima (Figure 23a). This area of over 28,800-nmi^2 offers a myriad of geologic features and possible tectonic processes because of the presence of a plateau with a horst, a fractured guyot, an obducted slab, the trenches, and an assortment of forearc seamounts.

At the northern portion of the complex, Uyeda Ridge (Figure 47) is either obducting or overprinting the trench. The sinuous ridge is the westernmost extant segment of the Chinook Fracture Zone. The ridge has such an influence on the forearc that the across-trench canyons appear to be an extension of the ridge.

Overall views (Figure 48) of the active margin onto the forearc show the convergence processes at work. There has to be a certain amount of compression deformation and faulting on both sides of the trench leading to the presence of several uplifted features. The horst is the most obvious. In fact, the horst may be

Figure 48. 3D views of the IBM margin from the NAVOCEANO multibeam sonar data bases, views provided by the GRASS package. The top image, from the south, is from 20°-25°N. The Ogasawara Plateau is on the top (bathymetry in Figure 47), the Udintsev Fracture Zone begins as the narrow ridge, and the Dutton Ridge is on the bottom. The northern Mariana forearc is characterized by a jumbled topography. The trench is 7-8 km deep. The central image, from the south, is 17°-21°N. The Dutton Ridge enters the convergent margin at the top as a series of unnamed seamounts lies to the south of that. After a bit of confusion in the bathymetry, the forearc takes on a more becalmed appearance. The active arc, on the west, is rife with active volcanoes. The bottom image is from 12°30'-17°30'N (bathymetry in Figure 25). Here the Magellan Seamounts are introduced to the convergent margin by whatever means. The forearc is deformed into what has been named the Lapulapu Ridge. Otherwise, it is rather straightforward as to the bathymetry until the active arc. There the islands Guam, Rota, Tinian, Saipan, and Anatahan lie from left to right.

the beginning stages of over-thrusting. Both the outer and inner trench walls are relatively clean for such an active region, although step-like, or terraced, features are on the outer forearc. The central slab of the plateau appears to have obducted onto the outer arc (see bathymetry on the cover). Samples show the slab to be ophiolitic, containing many rounded gravels of igneous rocks (harzburgite, dunite, gabbro, dolerite, and basalt predominate the 1000 dredge samples taken), which may be the product of inshore wave erosion. This Ogasawara "paleoland" is suggested to have been formed by forearc volcanism, a new production agent. The bathymetry shows the slab to have been obducted, the rounded cobbles having been produced by the insular erosion of the paleo-Broken-Top "Island." The horst on the southern part of the plateau, as well as the forearc canyons, are also a result of that stress.

Continuing onto the forearc, the Bonin Ridge would appear to be another of the outer arc compression ridges. Because the western slopes average 18% and the eastern 8%, tectonic control for its formation is suggested. The Bonin Ridge is presently not active, nor is it underlain by deep earthquakes. In fact, geophysical data indicate the ridge and the outer arc high to be the same structurally. However, it is not thought to be a compression ridge. The Bonin Ridge is a horst block associated with distributed rifting during the Oligocene. Here again, stresses from the Ogasawara Plateau could be the cause. The forearc north of here has been theorized to be a part of Mid-Oligocene rifting of an Eocene tholeiitic and boninitic arc massif. The cause of that rifting was proposed to be the arrival of the Ogasawara Plateau into the convergent margin by Adam Klaus and Brian Taylor of HIG.

The fault through Broken-Top Guyot gives its name. The back of the rear plateau wall appears nicely, with a possible fragment of the horst block still extant. The horst itself angles into the trench. The obducted slab now appears much the same as the tree trunk that has fallen across the stream and broken. So much of the old plateau is on the forearc that the seamount there is as large as Broken-Top Guyot. While the inner Izu Trench wall is relatively smooth similar to the Bonin Trench wall, the outer trench wall is another matter entirely. The fault through Broken-Top Guyot continues to the trench bottom. This is apparently the step like procedure for grinding up a large object to make it more easily digestible by the trench, so to speak. Last, the seamounts in the left foreground are on the northern outer flanks of the Michelson Ridge and may be parasite cones associated with the edifice building volcanism.

The upper subducting slab at the Ogasawara

Plateau is descending at a 45° dip angle for the first 150 km of depth. Once again, there is a break in recorded earthquake activity, where an almost vertical dip angle is described from 300-km on down. This "blue zone" is 350 km landward of the trench axis. The break was predicted by Benioff, and this is Benioff's proof, even as he predicted it would be 40 years ago. Directly above the vertical slab is a cluster of volcanoes.

The northern Mariana Trench is the scene of relative activity compared to that of the northern neighbor (Figure 48a). It bends sharply to the southeast as the active arc ridge bifurcates at 23°N latitude. The Ogasawara Plateau, an unnamed seamount, a NNW-SSE fracture zone ridge, and another seamount all lie seaward to the trench. No longer is the forearc a nice, sediment draped smooth region. There is a reason. The lineations imprinted on the landward lithosphere lead one to believe that they are formed by extensional processes, and that will be discussed in the tectonics section.

A 37-km diameter seamount at 22°N latitude may be only the tail end of a much larger subducting seamount or guyot being engulfed whose final legacy is to separate the Mariana and Bonin Trenches, echoing a similar fate as Kashima Seamount discussed above. The seamount has a downdropped portion, that portion having normal faulted 550-m. There is a 65-km long, 1830-m tall ridge on the inner trench wall that could be a re-entrant portion of the original edifice. A profile study give no hint, so even the small amount of stress being transferred by a glancing blow must be enough to form a compression ridge. There would necessarily be no effect on the earthquake or volcanic regimes, and that is just the case. The epicenters show many shallow earthquakes and no deep ones, the deepest being at 200-km. The dip angle here is 18°.

The cluster of volcanoes from 21.5°--22.5°N latitude is not only active but also growing: Eifuku, Daikoku, an unnamed one, North Eifuku, Kasuga, South Kasuga, and Fukujin. The "normal" active arc, the East Mariana Ridge, has a continuous line of volcanoes at 195-km from the trench. However, where large seamounts appear in the convergence zone, volcanism is increased to the point that each volcano becomes a cluster of volcanoes. In the sphere of influence for the four Magellan Seamounts the active arc between Aguijan Island and Pagan Island has 25 volcanoes. The extremely active "Lighthouse of the Pacific," Pagan, heads the string of five at Guguan, the cluster of five at Zealandia Bank, the three at both Medinilla and Anatahan, and the eight on the southern extreme. Esmeralda, Ruby, Guguan, and Pagan are the only active volcanoes on this segment, with Pagan erupting almost continuously since 1981 (from many Bulletins, Smithsonian Institution Staff).

The earthquake regime shows a lithosphere dip angle of 33°, with relatively few intermediate to deep earthquakes (Table III; Appendix B-3).

Table III. EARTHQUAKE DATA FOR THE IBM MARGIN

(Source is NGDC 1990 Earthquake CD-ROM)

Earthquake depth (kms)	Quantity
0-49	4994
50-99	1567
100-149	608
150-199	316
200-249	171
250-299	116
300-349	228
350-399	262
400-449	290
450-499	247
500-549	110
550-599	32
600-649	16
650-699	1
700+	0
unknown	243

As the Mariana Trench continues to the south, it appears to have migrated to the east from its past position. The Dutton Ridge (Figure 23b and 48b) occupies the seaward lithosphere between 20° to 21°N latitude. The northernmost 22-km diameter, 1110-m high seamount is obducting onto the inner trench wall. The central feature, Fryer Guyot, appears to be obducting the Mariana Trench. While the front presented to the trench is only 22-km wide, this feature is backed up by a sizeable "big-brother," the rest of the Dutton Ridge. Two fragments appear to be downdropped at least 925-m to fill the trench. A rather large seamount lies to the south.

Under the influence of the three closely spaced seamounts there should be a very deleterious effect on the active margin. The inner trench wall and the outer arc high are very complex bathymetrically. Horseshoe Seamount on the south is 111-km long, 74-km wide, and rises to a 1280-m peak. This is not a small forearc seamount. The primary lineations parallel the trench. South of here at 19°N latitude, two forearc seamounts were formed by serpentinite diapirism. A north-trending 37-km ridge comes off the north point of Horseshoe and ends on the inner trench wall. North of that point lies another group of outer arc seamounts with a trench-parallel low landward of the ridge. A major trend transecting the entire forearc parallels Horseshoe's NE-trending ridge. This trend appears to be a stress controlled fracture. Fractures like this are presumed to have been initiatory in forming the canyons and horsts to the north of here.

The earthquake regime shows a plate dip angle for the upper portion at 27° and 74° for the lower. Ranneft's transverse faults are apparent as two lineations striking ENE bisect the old Dutton "Plateau." These faults diverge by 6 to 7°.

Continuing to work southwards at this active margin, the two small seamounts at 18°30'N latitude and 18°N latitude are in the Mariana Trench. The northernmost seamount is 2745-m high and 40-km across and has filled the trench with 915-m of material. The southernmost seamount is 1830-m high and 22-km diameter and has filled the trench with 730-m of rubble. Based on the profiles, the large outer arc high across the trench attest to possible stress transfer by these small seamounts. The two small seamounts in such close proximity to each other are also enough to cause the deep earthquakes, but this is more likely a function of their fortuitous location. The active arc lies 340-km landward of the trench axis. The Benioff dip angle is 65° in this region. While that diverges from the proposed physical laws of nature, it nevertheless is based on actual earthquake evidence. The last deep earthquakes in this basin are adjacent to these two seamounts.

The southern portion of the Mariana Trench curves back to the west from about 16.5°N latitude. The Magellan Seamounts (Figure 23c and 48c) lie seaward of the Mariana Trench from 13°N latitude to 16°N latitude. The main body of the Magellan Seamounts, the Mariana Trench, and the forearc region between Aguijan and Pagan Islands on the East Mariana Ridge are shown. The seamounts are converging on the trench, headed by delCano on the north and continuing with Quesada, Victoria, and Espinosa seamounts on the south. Practical geomorphology dictates that all the larger seamounts are coalesced cones, and that the smaller never quite reached that stage. In the north, the seamounts also all appear to have formed on a 1500-m plateau. DelCano is about 55-km diameter on the plateau whose N--S diameter is 185-km and whose E--W dimension is undetermined. Its gross morphology is characterized by three large flank rift zones and one smaller trending northeast. The Mariana Trench is totally obliterated at 16°N latitude by the Magellan Seamounts at the exact difference as the heights of the larger seamounts, about 3660-m of rubble. The middle section shows the one portion of the seaward trench wall that is not cluttered with stress caused features. It is a gentle 4% slope showing a few normal fault lineations and a perched basin at 7100-m before the 7500-m trench axis.

Quesada Seamount, to the south of that, has apparently already undergone normal, or thrust, faulting twice. Profiles across that feature reveal terraces at 5500-, 4750-, and 3300- m. Quesada appears to have been about 35-km diameter and shows a similar geomorphology to other smaller diameter seamounts north of this point. Isolating Quesada Seamount in 3D demonstrates the downdropping.

The active margin includes Victoria Guyot. Both the north and south flank rift zones show signs of later volcanism. Gross morphology and the bisected top show Victoria Guyot to be coalesced cones. Continuing south the Mariana Trench shows one of its few rubble-free spots. The outer trench wall averages 8% slope and the inner averages 10%. The inner trench wall is marred by terraces, perched basins, and small bathymetric highs. The forearc itself averages 4% slopes and shows four 20--25-km diameter seamounts.

The southern portion shows the Mariana

Trench convergent margin at 13°N latitude as it bends west. Seaward, Serrao Guyot lies at the axis. The outer trench wall also slopes 20%. The trench axis has been shown to be constantly deepening, and the southern section is 9500-m deep. Offscraping is shown by three seamounts on the inner trench wall and one in the trench axis. Once again, the inner trench wall is easily studied because there is no accretionary prism. There are also several perched basins.

The forearc basin has the most bathymetric character from Japan south to the Caroline Ridge. Lapulapu Ridge appears to be an extension of the seaward ridge and rubble. In fact the Lapulapu Ridge is the largest forearc feature in this region. At 55-km wide and 115-km long it is larger than any of the approaching seamounts. The forearc seamounts form a "Z" shape with the lower leg of the "Z" being opposite the Quesada-unnamed seamount ridge and the middle opposite the gentle outer trench wall. A rather large depression west of the "Z" has been interpreted as being a canyon, but appears to be more of a terrace or basin which has been bulldozed upslope about 90-km.

The southern curve in the Mariana Trench is home to the deepest spot in the oceans, the Challenger Deep. The USNS MICHELSON got a depth of 10,915 m. The Japanese vessel TAKUYO, using SeaBeam corrected for sound velocity by CTDs, recorded depths of (1) 10,924 m at 11°22.4'N latitude, 142°35.5'E longitude, (2) 10,909 m at 11°20.2'N, 142°12.8'E, and (3) 10,901 m at 11°21.6'N latitude, 142°26.7'E longitude. They claimed the deepest, most accurate depth yet recorded.

In a place with so many of the characteristics to make for an excellent deep earthquake study, the earthquake regime shows a normal gradation from the surface to about 150-km, except for the region between 15° and 18°N latitude, where there is a long, narrow band of 200--350-km deep earthquakes, corresponding exactly with the active volcanoes. There are no deep earthquakes. The Benioff zone dip angle is generally 54° to 58°.

4. Yap Cusp

This cusp at the intersection of Yap (Figure 49-2) and Palau (Figure 49-1) Trenches is home to the Caroline Ridge. The ridge, thought to run from the trench cusp at 130°E longitude to Kusaie Island at 163°E longitude, actually loses its bathymetric signature at 146°E longitude, a distance of some 880-km. The ridge is incised parallel to the long axis by the Sorol Trough. The Sorol Trough was thought to be an inactive interarc basin, and the Caroline Ridge an extinct volcano arc. The Sorol Trough runs from Ulithi Atoll at 10°N latitude, 139.7°E longitude in an WNW--ESE direction to 5.9°N latitude, 145.4°E longitude. No sediments were found at that time. Fifteen years later, core samples were retrieved at the cusp which showed later reverse faults and kink bands. The later deformation is due to layer-parallel shear.

Earthquakewise, the Palau Trench is really a trough; it is seismically inactive. The Caroline Ridge has a resident earthquake population, and therein lies the key to unlocking the tectonic history of this section.

5. Philippine Sea Trenches

The Tosabae is a basement high at the trench slope break along the Nankai Trough, just to the southeast of Shikoku, Japan (Figure 49-6). The Kyushu-Palau Ridge (Figure 49-14) lies immediately to the west of the Tosabae, also on the Nankai Trough at 132°30'W longitude (Figure 49-6). The westernmost portion of the ridge is filling the trough. The ridge is causing a conspicuous depression on the inner trench wall. The forearc wedge and Toi Knoll are caused by the Kyushu-Palau Ridge. This is based on magnetic anomaly data.

The seismic stratigraphy which has been established for the Japan Trench is generally applicable here. In general, each acoustic unit has a uniform thickness with numerous unconformities. The combination of the great thickness (15 km or more) and uniformity across the are (with numerous unconformities) implies sedimentation in shallow water during a relatively stable tectonic regime. Unit I, which is interpreted to be Precambrian, has numerous coherent reflectors showing eroded folds (Appendix C). Unit Ia is deeply buried (more than 15-km below sealevel). Unit Ib is far thicker than its counterpart in the Japan Trench, where Ib is 2-km thick and truncated west of the trench. A structural high involving Unit I is located 30-km west of the

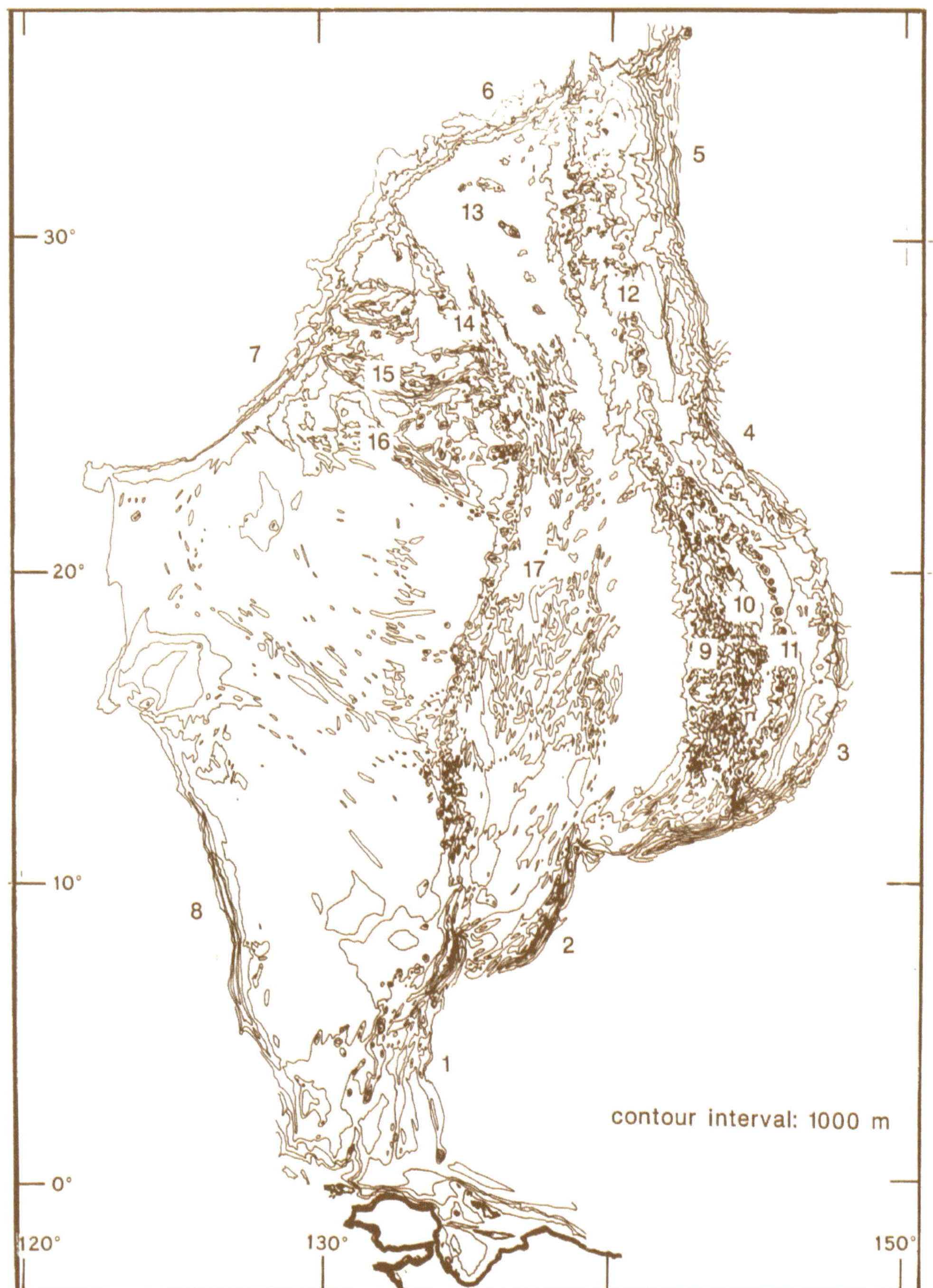

Figure 49. The Philippine Sea region at a 1000-m contour interval on a Mercator projection where: 1=Palau Trench, 2=Yap Trench, 3=Mariana Trench, 4=Bonin Trench, 5=Izu Trench, 6=Nankai Trough, 7=Ryukyu Trench, 8=Philippine Trench, 9=West Mariana Ridge, 10= Mariana Trough, 11=East Mariana Ridge, 12= South Honshu Ridge, 13=Shikoku Basin, 14=Kyushu-Palau Ridge, 15=Daito Ridge, 16=Oki-Daito Ridge, 17=Parece Vela Basin.

Nankai Trough, with an overall gentle, landward tilt. Unit I is considered to be orthoquartzite, particularly in the higher layers (Ic and Id) and to have supplied abundant orthoquartzite detritus to the Paleozoic sedimentary basins. These basins later eroded to provide material for the Mesozoic and Cenozoic formation on Shikoku Island. Unit II, interpreted to be Lower Paleozoic, shows NW progradation towards the present-day island arc. Unit III, interpreted to be Lower to Middle Paleozoic, onlaps southeastward toward the basement high west of the trough. Unit IV, Permian to Triassic, is developed only west of the trough in the area of the basement high and shows progradation toward the present-day island arc. Units V and VI thicken northwestward and bury structural lows of the underlying basement complex.

The presence of numerous, continuous, and coherent reflectors makes it possible to identify fault systems with relative accuracy. The crust is strongly fractured by well - developed reverse faults, tilted oceanward, that parallel the Nankai Trough. Most faulting seems to have ceased after the deposition of Unit V; that is, by Jurassic--Early Cretaceous. However, several large, deep-rooted faults are still active, disturbing the ocean floor bathymetry. The imbricate structure at the foot of the forearc slope, which has been claimed to be an accretionary prism by many authors, appears to be related to one of the large-scale, active reverse faults. The pronounced features developed in the crust, and the accompanying very thin sediment cover in the Nankai Trough alone, could account for the high heat-flow values. It explains heat flow more simply than the complex subduction mechanism. From the well-layered, folded, seismic stratigraphy, with its numerous erosional unconformities, together with the requirement of paleolands with orthoquartzite in the present Nankai Trough and Shikoku Basin (Figure 49-13) areas, the "oceanic" crust in these areas is more likely to be Precambrian continental crust.

The active arc is 150 km behind the trench axis, and the intermediate earthquakes lie 240-km beyond that. The earthquake regime is shallow, most of the activity taking place in the upper 200-km: 0-49 km deep region has 5245 events; 50-99 km deep region has 870 events; 100-149 km region has 282; 150-199 km has 132; 200-249 km has 43, and 250-299 km deep region has 12 events (Appendix B-3). The Benioff Zone dip angle is about 30°.

The axis at the bend of the Ryukyu Trench (Figure 49-7) is occupied by several ridges comprising the Amami Plateau between the Okinawa and Amami Islands. Minami Daito Ridge (Figure 49-15) is on the south at 27°N latitude, 130°E longitude. The ridge itself bifurcates, a low of 700 m deeper in that part of the edifice. Several 700 to 1000-m high seamounts reside on the outer trench wall before a prominent low at 27°50'N and 28°N latitude. On the north there are two WNW-ESE trending ridges with 1600--1800-m of relief that finish the Amami Plateau.

The Ryukyu Trench axis is 6800-m deep on the south. An 1800-m seamount at 27°N latitude in the axis is sitting on a relatively flat floor about 40-km across at 6400-m depth. The inner trench wall is characterized by steep protuberances bounded on the sides by lows. These are similar in geomorphology to the fluted walls of the Grand Canyon and are assumed to be erosional features. The forearc is a mixture of canyons on the south changing to compression folds on the north.

The active arc is 150 km behind the trench axis and is noted for volcanic activity associated with the Ryukyu Islands. The small clustering of earthquakes just to the north of Amami Island is presumably caused in part by compression as witnessed by the splintered ridge in the trench axis on the seaward lithosphere and the compressional ridge on the outer arc high. A series of earthquakes lies 250 km landward of the trench axis, most of them lying above the asthenosphere.

The Philippine Trench (Figure 49-8) presents an interesting case study from an earthquake and segmentation standpoint. Ranneft found a cross-trend with a magnetometer and field mapping through Mindanao. Called the Illana-Lianga Fracture Zone, it seems to be an area of major offset from the Philippine Trench on the east. The earthquake data verify this interpretation. Bathymetrically, a new trench appears to the west of Luzon. Because the central islands chain has been called "confused," that label is now removed by the presence of the transverse faults. All of the islands align with at least one of the major trends. Masbate parallels both.

94% of the earthquakes for this region are above the asthenosphere.

6. Equatorial Pacific Trenches

This large active margin lies to the southeast of the Philippine Sea basin (Figure 49). It is comprised of many trench segments having many names. The overall system (Figure 33) is called the Vityaz Trench System (VTS), and that system runs from New Guinea on the west to the Tonga Trench on the east. We will also lump the New Hebrides Trench in with this group, and the reason will be clarified: no apparent Vityaz Trench exists for much of the way!

Good survey data off northern New Guinea gives trench segment depths. The Manus Trench to the west of the intersection with the Mussau Trench has an axial depth of 6000. To the east, that depth lowers to 6500 m. The next segment is called the Kilinailau Trench, and it is 4500 m deep. Just to the south of that lies the New Britain Trench at 8000 m.

The OJP (Figure 39) may have been subjected to the northern migration of the VTS, or undergoing "trench capture." Reasonable speculation allows for compression ridges to be formed when the OJP and the VTS meet, with the exception of one small item. Earthquake activity at the active margins has been shown to increase with depth when large, aseismic, buoyant highs enter the sphere of influence of that type of margin. Should subduction be occurring, deep earthquakes would appear in the data set. In this case, they do not. This fact would lead one to deduce, then, that no component of movement to the south by the OJP is occurring. The earthquakes delineate a fault zone, not a subduction pattern. Also, the earthquake activity bottoms out at 300 km. Another problem exists with the updated bathymetry. A minimal amount of trenches is available for the subduction processes to occur. Because of this specifically, and because subduction below the asthenosphere in general has been repeatedly disproven by many authors, the conclusion is therefore reached that no subduction is occurring here either. Instead, this is a strike-slip zone, just like the Aleutian "trench." It is another of the large Pacific basin WNW--ESE-trending fractures.

Bougainville Guyot, at 16°S latitude and 166°45'E longitude and part of the d'Entrecasteaux Zone (Figure 33-5) at the New Hebrides Trench (Figure 33-4), presents an excellent study case because of it's extensive SeaBeam and submersible coverage. The guyot is a 48-km front parallel to the trench and 30-km wide. Its 3-km of relief makes it one of the smaller guyots. However, it is doing exactly what it is supposed to do at a convergent margin- wreaking havoc on the stress regime. Multi-channel seismic reflection data show a reef cap on top and a thick debris apron on the lower flanks. The early stages of guyot "digestion" are fully documented here where streamlining of the guyot is occurring. Several factors are responsible for this process, including mass wasting, tectonic erosion or offscraping, and the creation of a water-rich sediment layer between the guyot and the seaward lithosphere. No deformation of the leading guyot edge by normal faulting is occurring.

The effects on the forearc continue. The New Hebrides island arc is transected by many fractures. These fractures may be stress transfer related as is the one across from the Dutton Ridge. The major fracture transecting Espiritu Santo is a continuation of the d'Entrecasteaux Fracture Zone. The New Hebrides Trench has been bombarded with a steady stream of aseismic, buoyant highs. We have discussed the d'Entrecasteaux Ridge and Bougainville Guyot. The Malekula re-entrant lies to the south of that. The Efate re-entrant is 60-km south of that. This forearc feature lies landward of ORSTOM Seamount which, with a 33-km broadside, is still only a fraction of the re-entrant's size.

One possibility for the formation of active margin features is the formation of inner trench wall perched basins by re-entrant seamounts. At the New Hebrides convergent margin we are able to follow this process. The first step in the creation of a perched basin is the arrival of a seamount, in this case the Bougainville Guyot. The guyot has caused antiforms that include reflective, thrust faulted rocks. Various processes at work on both the guyot and the arc are bulk deformation of the guyot by sediment compaction and thrust faulting, tectonic erosion, and the accretion of water-rich, mobilized units of soft sediment. In the next step, seamount engulfment has formed an arcuate fold-thrust belt across the mouth of the re-entrant. This has formed the toe of a new accretionary wedge. At the Efate re-entrant the arcuate belt has grown by continued accretion at the toe of the wedge, underplating beneath the re-entrant, and trapping sediments.

At the New Hebrides Trench (Figure 33-4) region the volcanoes are again clustered, represented in this case by the leading arc of Espiritu Santo and Malekula and the trailing arc of Efate, Maewo, Ambrym, and the Shepherd Islands. They range up to 1879 feet high. Aoba Volcano on Ambae Island is the largest shield volcano in the New Hebrides arc. It lies at the end of the major fracture shown by Ranneft. The statistics on this, the most potentially dangerous of the Vanuatu volcanoes, are a base at 3000-m below sea level, a summit 1500-m above sea level, and a volume of 2500-km^3. This volcano is presently undergoing unusual seismicity, which is hazardous in a volcano with relatively young deposits, thick lahar deposits, the summit lake, and strong degassing.

ODP Leg 134 found data in the Vanuatu region where Indian-Australian lithosphere "subducts" Pacific lithosphere at a rate of 10 cm/yr to challenge the currently held models on convergence. Instead of the predicted series of prisms that allow seawater to percolate freely through the accretionary wedge, the site revealed a relatively uniform thickness of over one mile with few paths for water to ascend to the surface. A buoyant, aseismic ridge resides there.

Earthquake-wise, this region is a seismologists dream (Table IV; Appendix B-4). The composite of the Benioff Zone dip angle shows the intermediate earthquakes recording an angle of 70°. One study showed that, of 59 recorded events, only three were deeper than 240 km.

Using the 1990 NGDC earthquake tape gives 24,041 events for this region.

Table IV. EARTHQUAKE DATA FOR THE EQUATORIAL PACIFIC REGION.

(Source is the NGDC 1990 Earthquake CD-ROM)

Earthquake depth (kms)	Quantity
0-49	13665
50-99	6923
100-149	3350
150-199	1708
200-249	638
250-299	190
300-349	131
350-399	134
400-449	135
450-499	82
500-549	142
550-599	149
600-649	117
650-699	9
700+	0
unknown	845

7. Andaman-Java Trenches

This region has a scarcity of bathymetric data to formulate any kind of ocean floor tectonics hypothesis. The presence of trenches is known. A seamount resides on the seaward lithosphere at 10.9°S latitude and 109°E longitude. Its depth range is from 3070--4950-m. The Java Trench axis lies at 6970-m for this segment, jumping up to 6400-m to the east. A compression ridge across the trench rises up to 2000-m on the west and 1000-m on the east. The ocean floor is 1500-m below that on the forearc before it rises to the island arc. This small seamount, based on the DBDB-5, should not have such an effect on forearc and island arc events.

The Sunda Strait, separating Sumatra and Java, is typical for the region. The rift zones on the islands are offset, there is an abrupt change in strike between the islands, and there is a profusion of transverse faults. For Sumatra the Benioff zone is gentle, major longitudinal right-lateral transverse faults exist, and there is low volcanic activity. For Java there is a steep Benioff Zone, high volcanic activity, and its rift zone is nearly twice as wide as Sumatra's.

The active arc in the Indian Ocean basin is represented by the Indonesian Islands, which are home to many active volcanoes: Merapi, Slamet, Semeru, and Tengger on Java to be specific (Global Volcanic Network Bulletin). The active arc is 300-km landward of the trench axis. One of the most famous volcanoes in the world resides there in the Sunda Straits, Krakatau. Krakatau first collapsed in 416 A.D., leaving three islets on the caldera rim. The southernmost island, Rakata regrew as a basaltic cone, undergoing a series of eruptions. By 1680 A.D. two smaller andesite cones had coalesced with Rakata to form a 17-mi^2 island, which erupted in 1680 and left the island barren. In 1883 the craters on Perbuatan and Danan, the two cones, began ejecting ash. Five cubic miles of ash spewed forth and depleted the magma chamber. The volcano collapsed in on itself and created a caldera 4.3 miles in diameter. In 1927 the new island, Anak Krakatau, started growing in that caldera. "The Son of Krakatau" is active today. Continuing southeasterly, New Britain Island is host to Manam, Langila, Ulawun, and Rabaul volcanoes, all of which are hot. Ulawun has had 23 known eruptions since 1700.

The earthquakes for this region show a gap in seismicity from 350--500-km deep and another one beyond 650 km (Appendix B-2). The deep earthquakes are centered behind the active arc islands of Sumatra and Java, 460-km behind the trench axis: 0-49 km deep has 2853 events; 50-99 km deep has 1093 events; 100-149 km deep has 413 events; 150-199 km deep has 232 events; 200-249 km deep has 69 events; 250-299 km deep has 36 events; 300-349 km deep has 23 events; 350-399 km deep has 0 events; 400-449 km deep has 2 events; 450-499 km deep has 5 events; 500-549 km deep has 23 events; 550-599 km deep has 48 events; 600-649 km deep has 31 events; and 650-699km deep has 1 event.

Lastly, the trench system from the Andaman Islands to New Guinea has been in place since the Silurian-Devonian Era. It has been studied intensely and found to be the model for trench segmentation. A seismic profile exists across the Java Trench. The basement units (I and II) show vertical block movement, and are covered by a relatively undisturbed sedimentary layer (Unit III). Northwestward (present-day landward) sediment progradation is especially conspicuous throughout the overlying units (IV to VI), implying the presence of exposed land in the present deep ocean region when these units were deposited. Unit VII was deposited after the present oceanic region started to subside.

Thrust faults are seen in the lower slope affecting Units III to VII, and are usually attributed to plate subduction. However, since there are no compressional tectonic features in the underlying units, Choi thinks that they are better interpreted as the result of sediment overloading in the middle to upper slope. He argues that the block-faulted units (I and II), which indicate a predominantly tensional stress regime, and the well-layered, little-disturbed Unit III, are unequivocal evidence against plate subduction.

Due to the lack of drilling information from the area, it is impossible to assign definite ages to the acoustic units. However, the striking similarity between the Java Trench acoustic stratigraphy and that of Pacific Ocean margins suggests that Units I and II correspond to the Precambrian, Unit III to the Paleozoic, Units IV and V to the Paleozoic–Mesozoic, Unit VI to the Mesozoic–Paleogene, and Unit VII to the Cenozoic.

Choi argues that plate subduction below Indonesia also is contradicted by the following facts: shallow earthquakes are most intense under the nonvolcanic outer island arc, some 150 to 200 km away from the trench (where the "subducting" plate supposedly begins its descent); the Boxing Day quake main shock, too, took place under the outer arc, not under the trench (contrary to what some sources erroneously reported); there is a lack of seismic activity at around 400 km depth in the eastern Indonesian arc, and a lack of deep earthquakes (300 km or deeper) in northern Sumatra; the seismofocal zone is clearly controlled by a major deep-seated tectonic zone: the Shan Boundary–West Malaysia–Java Sea (SWJ) tectonic zone, which is visible in surface geological structures. Plate tectonics has failed to acknowledge that earthquake occurrence and distribution are firmly controlled by geological structures observed on the earth's surface.

8. Tonga Trench

The Tonga Trench (Figure 33-9), trending easterly, lies on the northern extreme of this system. Machias Seamount lies in close proximity, so it is considered together with the trench. The summit of Machias Seamount, at 500 fm, is 3500 fm above the trench axis. The feature, at 15°S latitude and 172° 15'W longitude, has northeast, northwest, and southwest flank rift zones. From SeaMARC II side-scan interpretations, a small fan lies at the canyon's end on the northeast slope of the trench. A large re-entrant is proposed to cut into the base of the guyot. The limited portion of the forearc is characterized by isolated hills, irregularly distributed ridges, and depressions. This 28-km diameter seamount should have no trouble being reduced to rubble as it descends to its ultimate demise, except that it has an escape clause.

The earthquake regime at Machias Seamount spills over on the seaward side of the trench, and herein lies the problem. For this seamount to enter the trench, the trapdoor spider analogy is appropriate. The Tonga Trench has been migrating northeastward for the past 2-Ma at an unknown rate. Should this phenomenon continue, the trench will eventually capture Machias Seamount.

The deeper earthquakes in the north of the Tonga Trench may have been caused by compression as the trench bends due to eastward growth by the backarc region (Appendix B-4). South of the major bend at 15°S a large feature, Peter Lonsdale of Scripps Institution of Oceanography (SIO) surveyed Capricorn Guyot at 18°S and determined that it is subducting. The volcanoes should be clustered on the volcano arc. The Tonga and Tofua Ridges are 300-km NW of Osbourne Guyot, and they should be the the active, volcano arc for this region. That is where the deeper earthquakes should reside; they do not. The Lau Ridge borders that to the west, 750-km landward of the trenches. There is a scarce amount of 300-km deep earthquakes below the Tonga and Tofua ridges. In fact, the clustered deep earthquakes, at 550-km landward of the trench axis, are under the Lau Basin.

Turning south, Capricorn Guyot at 18°35'S latitude and 172°20'W longitude does not show the outward morphology of compression deformation, probably because it is not close enough to have felt the insidious influence of those processes. The seaward lithosphere, however, is already under those influences and is buckling. Osbourne Guyot, at 26°S latitude and 175°W longitude, lies at the trench juncture is separated by a saddle, which on first glance appears to be a part of Osbourne Guyot. In actuality, the trenches are misaligned here just as Ranneft predicted. Osbourne Guyot itself has a tilted plateau of 275-km^2 areal extent. This shows that the feature is subject to compression stress already. The lithosphere around the guyot is broken into graben-like structures striking N--S. The inner trench wall has many terraces, basins, and general features associated with an aseismic, buoyant feature undergoing compression, showing a forearc, outer arc high, and inner trench wall that are incised by multiple canyons, leaving residual highs on the outer arc high.

The Tonga Trench axis is at 6500 m at Machias, and the Kermadec Trench (Figure 33-10) axis is at 7500 m at Osbourne Guyot.

The earthquake regime shows many deep events between 17°S and 27°S latitude. The deeper earthquakes cluster about 550 km landward of the trench axis. The upper Benioff Zone descends at 22°, and the lower at 70°.

Table V. EARTHQUAKE DATA FOR THE SW PACIFIC ACTIVE MARGINS

(Source is the NGDC 1990 Earthquake CD-ROM)

Earthquake depths (kms)	Quantity
0-49	11726
50-99	2543
100-149	1586
150-199	981
200-249	896
250-299	514
300-349	291
350-399	378
400-449	415
450-499	486
500-549	1208
550-599	1704
600-649	1112
650-699	196
700+	5
unknown	748

9. Peru-Chile Trench

Moving to the eastern side of the Pacific basin, the entire Peru-Chile trench margin is considered, primarily because of the lack of good bathymetry data for a micro-tectonic study. The Nazca Ridge comes off the East Pacific Rise. It affects the South American borderland in many ways. GLORIA surveys of the Peru Trench between 10°S and 14°S latitude show lithosphere broken by normal faults striking sub-parallel to the trench axis. The approaching lithosphere starts to bend about 100-km from the trench, where outcropping basement protrudes through the faults. This is analogous to the region around India where E--W ridges, whether close to the Himalayan foothills or as far away as the Indian Ocean basin, are attributed to collision compression. Seismic data show this to be the case. Rock samples are Precambrian in age. The trench is turbidite filled, with a depth of 6300-m at 10°S latitude rising to 5400-m at 14°S latitude. En echelon trench axis segments exist here.

Geophysics are used to help explain events in this data-poor section of the Pacific margin. The seismic profiles, along with gravity data, expand on much of what we have already seen for Pacific "subduction zones." Using the satellite altimetry map, one can easily follow lineaments from the Pacific ocean floor onto the South American continent.

A high-resolution seismic profile crosses the continental margin off Peru at 40°S latitude (Appendix C). The diagram shows no chaotic structural disturbances that one would expect at a subducting margin. A profile was reinterpreted without the outside influences of any preconceived notions about active margins. No DSDP/ODP drilling has taken place in this particular spot, so no ground-truth for the rock/Unit ages exists. However, and fortunately, another seismic line crosses the Peru Trench at about 10°S latitude. Choi has carefully traced the reflectors which are geologically significant.

Two ODP holes were made on this line, and three more holes are situated to the east of this profile. Stratigraphically, the continental margin in this area is divided into three major units: lower (Units I and II), middle (Units III-V), and upper (Eocene? and younger). Unit I shows clear folding structures cut by high-angle block faults, implying that it may consist of continental rocks. Unit II thinly covers Unit I and buries the depressions. The thick Unit III, well-layered and showing little structural disturbance, is widely distributed throughout the section. It appears to form ocean floor under the thin Cenozoic sediment layer beneath the trench axis. Unit IV also shows little tectonic deformation. A conspicuous landward progradation of sediment is seen in this unit, implying that the source area is on the west, an area now occupied by the deep ocean floor and the trench. Zuniga-Revero and others (1999) work emphasized the presence of thick sedimentary layers beneath the Paleogene unit based on the exploratory well results and the modern seismic data. A comparison between Choi's and Zuniga-Revero and others' interpretations suggests that Units IV and V can be correlated to the Cretaceous, and Unit III to Paleogene in age.

Off the western shore of South America, we find yet more of the same. From a tectonic point of view, the area has had little disturbance and it has stayed relatively stable after the deposition of Unit I and subsequent folding and block movement. Except for several minor thrusts at the foot of the continental slope near the trench, no major thrusts represented by the underplating of oceanic crust is present. A series of seaward-dipping normal (active and growth) faults that disturb the upper (Eocene? and younger) sediments is well-developed at regular intervals. One of them between ODP Sites 653 and 685 may be responsible for forming the paleoslope before deposition of the middle Eocene? slope sediments. The presence of paleoland during the Paleozoic to Mesozoic in the SE Pacific near the Peru-Chile Trench is in accord with the above seismic interpretation. This leads one to believe that paleoland is still present, although it is submerged to the west of the trench.

Stratigraphically the continental margin of the southern portion of South America is divided into three major units; lower (Units I and II), middle (Units III to V), and upper units (Eocene? and younger). Unit I shows clear folding structures cut by high angle block faults, implying it may consists of continental rocks. Unit II thinly covers the Unit I and buries depressions. The thick Unit III, well layered and showing little structural disturbance, widely distributes throughout the section. In the trench it appears to form sea floor underneath the thin late Cenozoic sedimentary cover. Unit IV also shows little tectonic deformation. A conspicuous landward sediment progradation is seen in this unit (marked by open arrows), implying that the source area is in the west of this area, now occupied by the trench and deep sea. The presence of thick sedimentary layers beneath Paleogene unit based on the exploratory well results and modern seismic

Figure 50. Dong Choi provides this diagram of the rock ages in South America with oceanic megatrends passing through the continent. This map, coupled with several re-investigated seismic stratigraphy profiles, shows the fallacy of assuming that continents and oceans are not mutually exclusive in age.

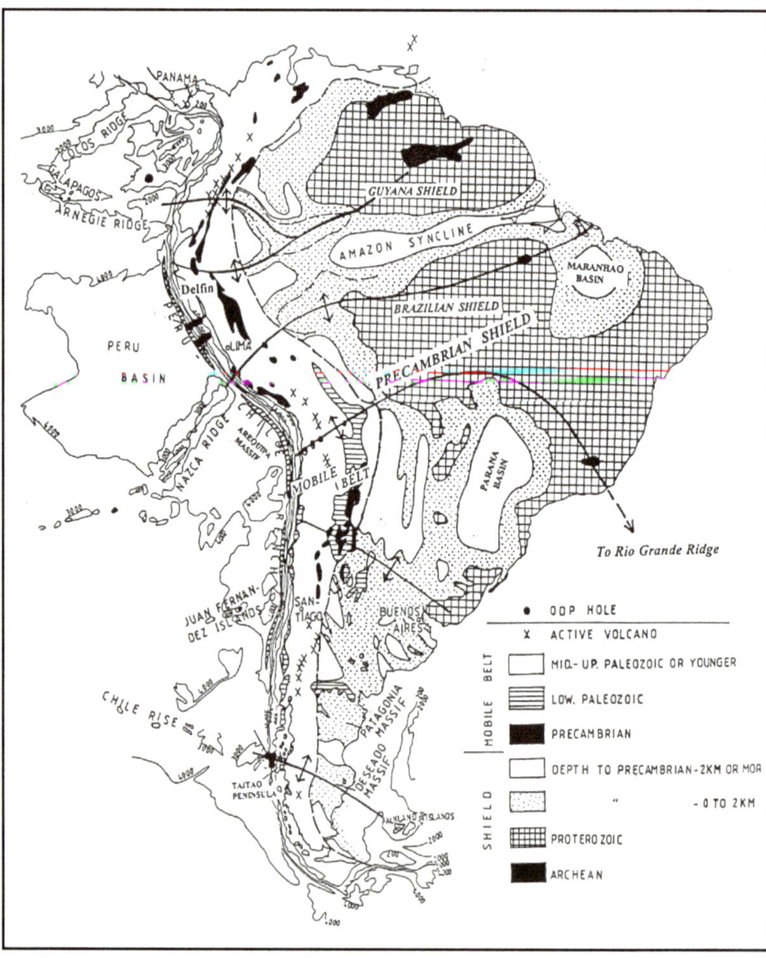

data is emphasized. Comparison between the two seismic interpretations suggests that Units IV and V can be correlated to Cretaceous and Unit III to Paleozoic in age.

The seismic stratigraphy of the Peru continental margin shows close similarity to that of the Japan and Aleutian Trenches (see Appendix C). Unit I in all areas is block faulted and directly underlies thin younger (Mesozoic to Cenozoic) sedimentary cover in the present trench and deep sea. Its structural (geanticlinal) high is located in the present trench and deep ocean areas, dipping gently landward. In the Japan and Kuril-Kamchatka Trenches, Precambrian continental rocks have been dredged from Unit I and/or II. Unit II thinly covers and buries lows formed by Unit I, and shows a limited distribution. Units III and IV are thick and wide with Unit III showing a strong coherent layering. In all of these areas, judging from sediment progradation pattern, the provenance for the Units III, IV and V sediments is considered to be in the present deep oceans.

New geological interpretation of the crustal structures based on seismic data interpretation in all trenches examined here obviously show no signs of subduction. The "oceanic crust" continues under the continental slope without subduction. The trench is situated near the axis of the basement (lower crustal) high which is intensively faulted and eroded. Moreover, the dredging results indicate that the "oceanic crust" consists of Precambrian and younger continental rocks in the NW Pacific. Their decollements coincide with either unconformities or thrusts.

Studies have analyzed the spatial distribution of the earthquake regime for this area, finding that there are five distinct segments of inclined Benioff Zones (see Appendix B-7). Two of the segments from 2° to 15°S latitude and from 27° to 33°S latitude have 10° dip angles. Three segments from 0°S to 2°S latitude, from 15°S to 27°S latitude, and from 33°S to 45°S latitude have steeper dip angles of 25° to 30°. Most of the seismic activity is in the upper 50-km of the landward lithosphere. In the flatter lithosphere-dip regions there is no Quaternary volcanism. They also noticed a gap in seismicity between 320 and 525 km depth. The NGDC earthquake regimes also show a marked decrease in earthquake activity between 320 and 525-km. Interestingly, two lines of deep earthquakes lie under South America, one from 6 to 11.5°S latitude and one from 19 to 29°S latitude. The northern one corresponds to a shallow dip angle, and the southern one corresponds to a steep dip angle.

The earthquake hypocenters are densest under the middle continental slope between the trench and the coast, around 80--150-kn east of the Peru Trench. The depth range is generally 25--70-km. Shallow to very shallow earthquakes are located in this area, too, and only three hypocenters are scattered under the trench at about 25--30-km into the mantle. In addition, no earthquakes have been recorded along the alleged megathrust or on the upper surface of the "underplating" oceanic crust. The seismically active zone is not located at the trench, but about 80--150-km landward of the axis.

The dip angles correspond to the location of the large aseismic, buoyant highs in the active margin that lie on the landward lithosphere. The Carnegie Ridge, the Nazca Ridge, and the Juan Fernandez Ridge all continue onto the South American craton (Figure 50). This ties in with the earthquake distribution and clustered volcanoes previously noted in the NW Pacific, although only 253 earthquakes out of 19,925 being recorded deeper than 300 km. A gap in activity essentially occurs between 300--500-km).

On the hypocenter plots of the earthquakes at certain margins, the deep earthquake events all lie in a region of supposed deep fault zones, they are generally stacked, and they are delineated on the surface by tectonic belts. For the western Pacific this means trenches and active arcs. For the eastern Pacific this means uplifted features such as the Andes Mountains.

Appendix B includes the Peru and Chile trenches off the west coast of South America, giving depth to Choi's study. In a series of six lines across the trenches, he found that almost all of the earthquake activity was above 200-km. It also shows a variability of descent angle, intermingled of 12°, 19°, 12°, 17°, 27°, and 20°. This is a really crooked plate! While the epicenter portion of the original figures investigated by Choi show deep earthquakes, the size of the dots precludes a real understanding of events there. In all six of the original lines, no events are noted between 200 and 600-km. In other words, no intermediate events occurred. No continuously descending slab exists, and all of the deep quakes are clustered below the Western Brazilian Tectonic Zone. This is a region from 5°S to around 27°S in two of the nine available spots worldwide for deep earthquakes to appear.

Choi concludes that deep earthquakes are related to deep tectonic zones which are recognizable on land and that the deep earthquakes occur in subsiding areas. They are directly responsible for the subsidence of the upper mantle and crust along the major deep fault systems, such as the western Pacific and the Peru and Chaco-Panana Basins off South America. He has found that the seismic focal plane leans oceanward, which indicates that the Benioff zone is a reverse thrust fault system.

The axes for the Peru Trench are 7000 m south of the Nazca Ridge and 5300 m north of the ridge.

The conclusion here is based on previous facts presented plus the additional data. The seismic data across the trench and continental shelf supported by exploratory drilling indicate the presence of thick Mesozoic (acoustic Units IV and V) and possibly Paleozoic (Unit III) sedimentary units underneath the Paleogene sediments. These are underlain by block-faulted, layered basal unit (Units I and II) that crops out or is covered by the basaltic layer in the trench and the SE Pacific. The sediments is Unit IV show a clear progradational pattern toward land, which indicates that the provenance of Unit IV was located in the area presently occupied by the deep ocean. This fact supports many previous geological studies made on the South American continent which assumed that the paleolands were in the Peru-Chile area during the Paleozoic and Mesozoic times . Obviously the Nazca "plate" has not subducted. The distribution of the earthquakes together with the above mentioned geological and geophysical studies negates the application of the plate-tectonic model to this region. A continental-scale paleoland existed in this region, and it supplied sediments to the geosynclinal basins which developed in the Pacific mobile belt along the coast. They have submerged during the Late Mesozoic and Tertiary.

The situation is similar to the Aleutian Trench, Japan Trench, and Kuril-Kamchatka Trench. The available evidence, 1) presumed paleolands, 2) continuation of continental Precambrian structures into the ocean floor, 3) new seismic profile interpretations as shown by Choi, and 4) the above mentioned Benioff zone and crustal structure in the Peru Trench area, would indicate that plate subduction has not taken place along the Peru and other trenches in the Pacific.

10. Middle America Trench

This is the site of many interesting geological events and possibilities (Figure 38). To begin with, the Cocos Ridge is really three ridges atop a plateau converging on the Middle America Trench at 8°N latitude and 83 to 84°W longitude, striking 055°. The northernmost ridge is at least 240 km long, 18 km wide, and has 1000 to 1400 m of relief above the regional depth of 3000 m to the north. On a 2000 m plateau, the central ridge is 90 km long and rises to a 600 m summit. The southernmost ridge is 240-km long, rising to 1000 m. The entire front is 180-km wide.

The Tehuantepec Fracture Zone fronts this trench. The Tehuantepec Ridge front is 55-km wide and rises about 1000-m over the surrounding ocean floor at the trenches. The long feature strikes 042°. This ridge seems to have no effect on the forearc, the bathymetry of which is from NGDC ship-of-opportunity data. It appears to downwarp with the surrounding region into the trenches. The trench axis is obliterated by the incoming fracture zones at 9.5°N latitude between 82°W and 83°W longitude. The NGDC bathymetry, while presenting an excellent regional field of coverage, does not show events on the inner trench wall or forearc other than to show a steep trench slope. The Cocos Ridge has overridden the trench. From the east, the Polochic Fault extends westward from Guatemala along 15°15'N latitude to offset the Middle America Trench by about 130-km. The northern trench axis is generally about 5400-m deep, and the southern trench axis is generally between 6400--6600-m.

The active arc at Central America is extremely virulent: Poas, Arenal, Irazu, Masaga, Momotombo, and Telica volcanoes. All of these have constant fumerolic rumblings. Arenal began erupting in 1968 in an unbroken sequence of Strombolian explosions and basaltic-andesite discharges from multiple vents. Poas is the most active, reaching an average of approximately 220 earthquake events per day in 1993. The summit contains several eroded calderas and two intermittent lakes. Many high, and highly active, volcanoes lie to the NW (Neo-volcanic Plateau) and SE (Central American Ranges), but none here. The Central American Ranges are also offset by the Polochic Fault, and the timing on that event is at 15 Ma.

Published DSDP seismic lines cross the trench (DSDP Leg 67]; ODP Leg 84). A migrated depth section of Line GUA-18 was the most intensively studied here (Appendix C). Most of the DSDP holes lined up along this transect and were used for control in the interpretation. For the sake of stratigraphic consideration, a geological profile through the Polochic Fault in Guatemala is also cited. The region is characterized by a faintly layered and folded mound-forming lower unit (in the lower slope, trench, and deep ocean it is labeled as Ia and Id) and a coherently and well-layered sedimentary cover which is well developed in the middle and upper continental slope. From the comparison with drilling results and nearby land geology in Guatemala and Mexico by the North American Geologic Map Committee (1965), the lower unit is considered the Pre-Devonian or Precambrian to Early Paleozoic basement, with the sedimentary cover ranging from Late Paleozoic to Cenozoic.

The Pre-Devonian basement (Unit 1) has the lower subunit, which shows a relatively well-layered and folded structure (Ia) covered by a thick and faintly layered and folded subunit (over 7-km thick; Unit Ib) with two thin top units (Ic and Id). A remarkable erosional unconformity is detected between the Ib and Ic units. Two holes, 269 and 267, reached the top of these units, Ic and Ib respectively. The former recovered amphibolite (altered gabbro and diabase), and the latter sheared and metamorphosed mafic rocks (metagabbro, metabasalt, and serpentinite mud) beneath the Upper Cretaceous limestone associated with serpentinite.

Unit II exists under the trench as a well-layered and folded unit. It is acoustically similar to the unit at the bottom of the section in the north of the horst (middle to upper slope, Unit II). It is bounded by the above basement units on both sides. We consider these units to be correlated with the Upper Paleozoic (Carboniferous to Permian) layers in the north Polochic Fault, Alta Verapaz, Guatemala judging from the acoustic character, folding, and stratigraphic relationship (angular unconformity) with the overlying units. However, the structural disturbance of Unit II under the middle slope appears more intense than that is Alta Verapaz, and therefore, the assignment of Unit II is still tentative.

The well-stratified, northward-sloping units above Unit II in the middle and upper slope are considered to be Mesozoic and Cenozoic in age (Jurassic, Cretaceous to Oligocene). They show downlapping northward (present land) and onlapping southward against the basement complex, implying that the basement had been subaerial in the present trench area.

Structurally, the present trench is located near the axis of a large geanticlinal structure of the Precambrian-Lower Paleozoic horst-block. The area had been deeply eroded (subaerially) before the deposition of the well-stratified Jurassic-Cretaceous-Paleogene sedimentary sequences, supplying sediments to the basin in the north, or the present Central America.

The surface unit (Neogene to Quaternary; sequences 1, 2, and 3 by Ladd and Schroder, 1985) was deposited after the paleoland, which had existed in the trench area, completely submerged under water. DSDP Sites 567 and 566 revealed the missing Paleogene sections, or the unconformity beneath the Miocene. This implies that the present trench and lower slope were subaerial during the Paleogene.

Along the Pacific coasts of the Central America several ophiolite blocks have been known. The presence of mafic-ultramafic metamorphic rock block in the continental slope off Guatemala is another addition to this linear distribution of oceanic crust. The presence of this basement block is also detected in a seismic profile (GUA-4) taken about 120 km west of the GUA-18 line, although its detailed interpretation is impossible due to the unavailability of migrated depth or time sections. Its fault boundary with the sedimentary layers is situated at the shoulder of the continental slope.

Thus, the continental slope off Guatemala is characterized by the basement horst that continues under the deep ocean floor of the East

Pacific where it is covered by Mesozoic and Cenozoic basaltic rocks. The trench coincides with an axis of geanticlinal basement high that was subjected to long-lasting subaerial erosion coupled with intensive faulting along the axis and resulting collapsed and incised structure.

Noteworthy blocks of Pre-Devonian metamorphic rocks, which are comparable with Unit I in the Middle America Trench, are distributed along the deep-seated Polochic-Motagua tectonic zone. This fault is an extension of the Cayman Trough(Figure 3-38) and the Clipperton Fracture Zone (Figure 15-11), a fact which warrants the conclusion that the mafic and ultramafic basement is associated with major tectonic zones in Central America, and that the basement horst is uplifted Precambrian--Lower Paleozoic massif. This is partly the real composition of the "oceanic" crust under the ocean floor elsewhere. The Middle America Trench is formed on or near the axis of Precambrian basement highs. In all of the trenches examined herein, the trenches are situated near the axis of an anticlinal structure of the lower crust (at least partly Precambrian in age composed of mafic and continental rocks) that has been subjected to intensive vertical tectonics as well as long-lasting subaerial erosion before submerging to the present depth.

This new analysis revealed that the continental slope off Guatemala is characterized by the basement horst that continues under the deep sea floor of the East Pacific where it is covered by Mesozoic and Cenozoic basaltic rocks. The trench coincides with an axis of geanticlinal basement high that subjected to the long lasting subaerial erosion, coupled with intensive faulting along the axis and resulting collapsed and incised structure.

No deep earthquakes exist here (Appendix B-6). Landward of the slope, a few intermediate earthquakes have been produced around 11°N latitude and 86°W longitude: 0-49 km region has 4602 events; 50-99 km deep region has 1690 events; 100-149 km deep region has 500 events; 150-199 km region has 228 events; the 200-249 km deep region has only 12 events. No cluster of earthquakes northeast of the trenches delineates any deleterious effects by the Tehuantepec Ridge. The Benioff Zone dip angle is 45°. The ridge is on a direct line with the Mexican city of Villahermosa, which is the other side of an isthmus. There are no high, or highly active, volcanoes here. The Central American Ranges are also offset by the Polochic Fault, the timing on that event at 15 Ma.

The axis for the Middle America Trench is 6000 m SE of the Tehuantepec Ridge and 5000 m NW of the ridge.

11. North Pacific "Trench"/Strike-slip Zone

The Cascadia zone off the northwestern United States (Figure 51) presents a very interesting case. Much has been written about the volcanoes because of their activity and proximity to human population centers. All of the Cascade volcanoes were built atop older material, the present pulse of volcanism beginning during the late Eocene and Oligocene. Magma floods from the Miocene Columbia River episode partially surrounded and covered many of the incipient Cascades. Folding occurred on the western Cascades, which was followed by more volcanic activity. During the late Pliocene and Pleistocene the volcanic activity moved to the east to build a parallel ridge. The most characteristic lava type on these stratovolcanoes is andesite, which flows for shorter distances and builds steeper volcanoes. The Cascades have historically produced all types of eruptions. Cataclysmic tephra eruptions have occurred at Glacier Peak, Crater Lake, and St. Helens. Mazama (Crater Lake) also underwent a Plinian eruption to produce Crater Lake. Pyroclastic Pelean activity occurred at Hood and Shasta. Lassen Peak, at 10,000' in northern California, is the southernmost of the Cascade volcanoes. Its last eruption was from 1914 to 1917. The northernmost, 8800' dacite peaked Garibaldi, has not erupted historically.

Mt. Hood was active from 1760-1810, 1859, and 1865 which reduced its 12,000' peak to its present height of 11,245'. Later activity on Mt. Adams has built that edifice into a composite cone. Mt. Rainier reaches the almost ultimate height for the Cascade Mountains at 14,410'. The peak was really about 16,000', but erosion by glacial action has reduced the size. The peak, or side vent, is called Tahama. St. Helens was most recently active in the 1980s, the Plinian eruption reducing its 9677' cone to less than 8000'. As the volcano weathers, the edifice is

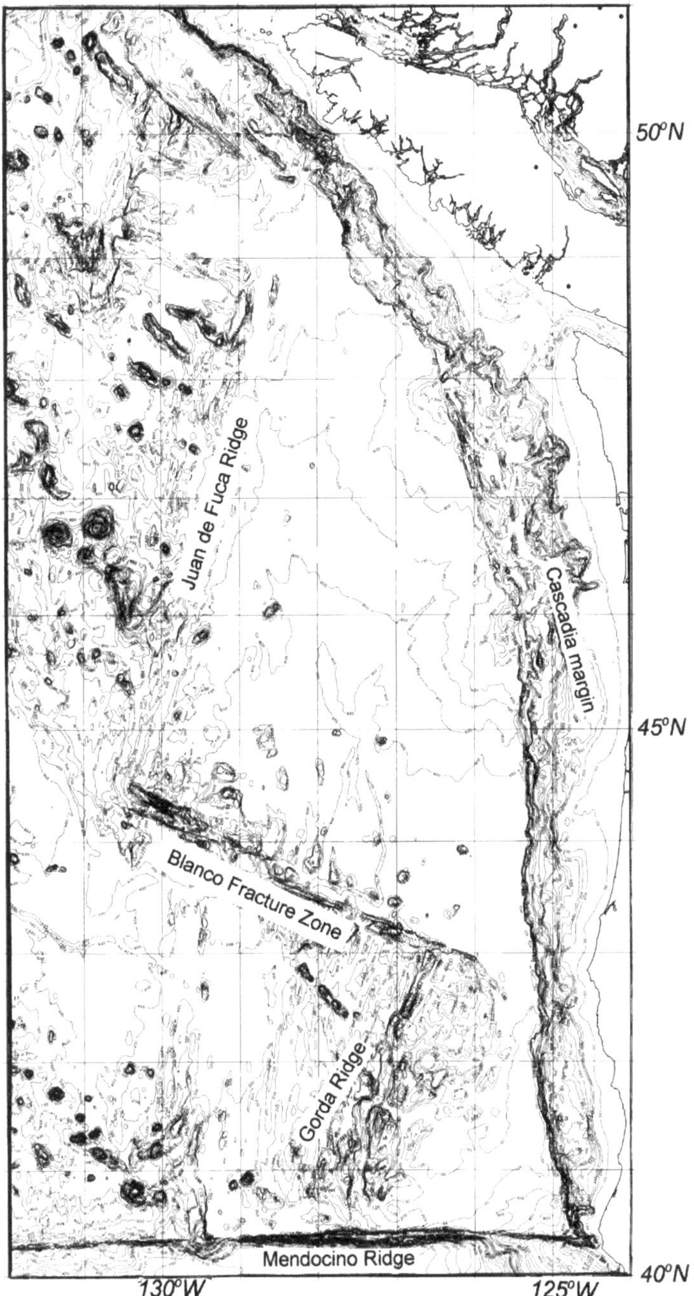

Figure 51. The Juan de Fuca and Gorda "plates" at a 100-fm contour interval on a Mercator projection. This eastern Gulf of Alaska section shows several of the trends. Of interest here also is the lack of a trench at the Cascadia margin. The shelf and slope plunge directly down to the plain. A look at the earthquake information in the Appendix also shows no activity in this region, leading one to wonder why this region was ever listed as a subduction zone.

left to become a volcanic pipe and, ultimately, a flattened peak such as at Goat Rocks.

This is all good and well. However, for this region no trench exists in the bathymetry (Figure 51) or the seismic traces, and most of the earthquake activity is above the asthenosphere where it exists at all (Appendix B-5). In fact, it parallels the San Andreas transform fault. However, the active arc is indicative of an active margin, and this region is historically included in summaries of "subduction zones."

The Gorda and Juan de Fuca ridges are called the spreading centers for the Gulf of Alaska. Offset by the Blanco Fracture Zone, they present yet another anomalous fly-in-the-ointment. They are oriented SSW-NNE. Seamount provinces formed on-ridge at the Juan de Fuca Ridge are all perpendicular to the spreading axis, as proposed. However, they are not parallel to any other seamount provinces in the Gulf. Also, they only grow off to the west of the spreading ridge. To the east of the ridges lie the proposed Juan de Fuca and Gorda plates. As no subduction zone exists to the east and no apparent spreading indicators, one has to wonder why these are called "plates."

Continuing to the west, the small, 18-km diameter Kodiak Seamount, at 57°N latitude and 149°W longitude, lies seaward of the Aleutian Trench. The seamount has several flank rift zones, but it has not begun to break up.

For this convergent margin there is a rather large portion of available SeaBeam data, most covering the trench axis and the landward side. It deepens from 5800-m on the east at 158°W longitude to 6900-m on the west at 163°40'W longitude with no sign of segmentation.

Derickson Seamount is a small seamount in the Gulf of Alaska at 161°W longitude at the Aleutian Trench. The size seems insignificant at 37-km long, 18-km wide, and rising not quite 1830-m over the ocean floor. It is still some few miles from the Aleutian Trench axis. However, the effect is much larger than would appear. The inner trench wall and outer arc high are marked by small canyons, perched basins, and Unimak Seamount. This is a region of maximum stress, so the features are directly related to this. Unimak Seamount's major axis is 49-km and parallels the trench. With a minor axis of 22-km, Unimak is larger than Derickson. Otherwise, the inner trench wall is laced with thrust faults, oblique-slip faults, and anticlines.

The forearc at Derickson Seamount is incised by at least two canyons which circumvent the edges of Unimak Seamount. These canyons are several hundred fathoms deep and have V-shaped bottoms, meaning they are probably still active. Between Unimak Seamount and Unimak Island, the active arc, there is a nondescript 73-km wide shelf.

The active arc north of the Aleutian Trench lies 95 km north of the trench axis. Directly above a large cluster of earthquakes centered at 60°N latitude and 153°W longitude, the volcanic field lies in the southwest Cook Inlet region. Most notable of the active volcanoes are the dome-clustered Augustine at 1282-m and the stratovolcano Redoubt at 3108-m. Augustine erupted most recently in 1986, and Redoubt erupted most recently in 1989 to interfere with airline traffic. Redoubt, in fact, has recorded at least 30 Holocene eruptions.

The next large cluster is near Unimak Seamount. In the small segment of the active arc there are 16 volcanoes lying 213 km from the trench axis. Bogoslaf Island lies 92 km behind the main volcanic front. A 1796 eruption built Castle Rock, and Bogoslaf has erupted again in 1804, 1806, 1883, 1906, 1913, 1926, 1931, and 1951. The front range consists of Makushin, Akutan, Lava Peak and Lava Point on Unalaska Island, the extinct Gilbert on Akun Island, and Unimak Island. This 120 km long island is home to ten basaltic-andesitc 6000 ft stratovolcanoes which are roughly Plio-Pleistocene in age.

All of the earthquakes in this region are above the asthenosphere: from 0-49 km there are 8254; from 50-99 km are located 1947; from 100-149 km are 847; from 150-199 km deep are 162 events; from 200-249 km are 69 events, and from 250-299 km are 5 events. None exist deeper than that.

A reinterpretation of the seismic profiles across the Aleutian Trench (Appendix C) has allowed an updated scenario for that region. The validity of the current working hypothesis was under close scrutiny for this region very early because of the undeformed trench sediments. McCarty and Scholl (1985) attempted to demonstrate the presence of subduction-related decollement thrusts and subducting sediment blocks. Large blocks of fresh amphibolite schist and hard graywacke are regarded as in situ at 173°-174°W longitude. Later use of a seismic profiler found the blocks to be truly in situ and aged them as Paleozoic or Mesozoic formations cropping out on the lower slope fault scarp.

In six unconformity-bounded units, each has several minor unconformity- bounded subunits. Out of the three examined, two profiles were reinterpreted. Unit I, acoustic basement, forms a positive mound whose apex is beneath the Alaska Abyssal Plain. Although inconspicuous, on a more thorough scrutiny, wavy, discontinuous, high-amplitude, low-frequency reflectors are traceable. The lower boundary, or Moho, is not clearly determined in the profiles due to the lack of velocity information. However, a relatively continuous series of reflectors at about 12 km below sealevel could perhaps be regarded as that. The layering in Unit I becomes more conspicuous the further up in that layer one looks. An extensive erosional unconformity accompanied by tectonic disturbance is discernible in the middle of Unit I. Unit II buries the bathymetric lows of Unit I, and the upper boundary is clearly erosional and clino-unconformable. This implies a tectonic disturbance before the deposition of Unit III. Unit III shows southward overlapping against the basement high, which suggests that the area of provenance is to the south; that is, oceanward of the trench. Unit III is well layered with little deformation, and it fills the lows of the underlying Units. Several subunits, each bounded by unconformities, are recognizable. Locally, Unit III progrades landward, filling the bathymetric lows beneath the landward portions of the trench slope. The lack of structural disturbance in this Unit is a strong evidence of regional stability after the deposition of Unit III.

At this time a substantial change in sedimentary environments seems to have taken place. The area of deposition for Unit IV has moved landward, and it has become a deposit of "geosynclinal" character. This assumption is based on the thick (3 km or more) and wedge-like character of the sediments in the profiles. Clear sediment progradation toward land is now obvious in this layer, as is also the case in the underlying Units II and III. Unit V overlays Unit IV with erosional unconformity. It is a 2-km or more thick sequence beneath the slope, which thins toward the abyssal plain. In the seismic profiles, its sediments show landward progradation. The top of this Unit is deeply eroded, particularly under the middle slope. This forms a trench-parallel large depression. Unit VI buries all of the above, and the top is the present ocean floor. The sediments were shed from the present island arc.

Correlation of the above acoustic units with established stratigraphic and lithologic units is hindered by the lack of available borehole information along the seismic lines. DSDP Sites 186 and 187 do not provide sub-bottom information. However, dredge haul information fills that bill nicely. The upper slope dredges recovered a homogeneous assemblage of diatomaceous tuffaceous siltstone of Pliocene age (A37D, between 5100--5500-m deep). Pliocene to Pleistocene rocks were also recovered. The sedimentary rocks are considered to have come from Unit VI. A deeper haul (A38D, between 6000--7000-m deep) obtained a very heterogeneous assemblage of igneous, metamorphic, and sedimentary rocks. A large, fresh block of amphibolite schist from this assemblage was originally considered to be possible basement. Gravity data are not supportive of this speculation. The comparison between the dredge site and the reinterpreted seismic profile indicates that the amphibolite and possibly greywacke have come from Units IV and/or V.

The "trench" is really not a "subduction zone" trench at all; it is a strike-slip zone. No component of convergence exists here according to the stress field maps. Most of the earthquakes are stacked in a vertical arrangement. And, the "trench" is segmented at regular intervals, so different forces seem to be at work here. From 164°W to 161°W longitude the axial deep is 6800-m deep. From 156°W to 153°W longitude the axial deep is 5800-m deep.

12. Caribbean and Scotia arcs

Moving across to the Atlantic basin, South America is framed by what appears to be flow structures, and they are listed as the Caribbean and Scotia plates. The Atlantic basin is theorized to be moving to the NW. The Caribbean and Scotia regions are moving eastward. Stress is being produced. Stress on the south is relieved by normal faulting, and the stress release on the north is strike-slip. DeMets and others also show the "plate boundary" between the North and South American "plates" to intersect at the SE limit of the Puerto Rican Trench at 17°N latitude. There is neither bathymetric nor geophysical evidence for this arbitrary placement of the boundary. In fact, the boundary has been variously placed at 15°N latitude and 19°N latitude.

The Caribbean region is an extreme case of convergence where 90 Ma lithosphere is converging at only 2 cm/yr. The convergence rate has been found to be to be 0.7 to 1.5 cm/yr. A study of 17 earthquakes showed no thrust faulting, so the conclusion was that the lithosphere boundary is decoupled, there is aseismic subduction, and that there is no interplate deformation occurring. The conclusion here is that no plate boundaries on either the north or the south exist, so an alternate model for the evolution of the Caribbean region is necessary.

With the predicted lack of stress on the northern boundary, no reason exists for an active arc in the Caribbean region. However, many extremely active volcanoes are in this region on the Antilles arc. Mt. Pelee is on Martinique, La Soufriere on St. Vincent, and Kick-'em-Jenny off Granada are the most noteworthy. All are between 12° and 15°N latitude. The active arc is segmented, showing a NE-trending transverse fault north of Guadalupe and La Desiderade Islands, an eastward-trending transverse fault between Martinique and St. Lucia, and a SE-trending transverse fault south of Granada. These are compounded by the effects of the St. Lucia-Barbados Crosswarp and are probably responsible for the intense volcanic activity in the region rather than the subduction process itself. In fact, surprisingly few earthquakes exist.

The earthquake regime for the Caribbean region: 0-49 km deep has 3308 events; 50-99 km has 442 events; 100-149 km has 236 events; 150-199 km deep has 98 events; and 200-249 km deep has 4 events. No earthquakes exist in this region below that depth.

Hence, the Puerto Rican "Trench" is a theoretical subduction zone, and it appears to have been squeezed out between North and South America. There are no large, aseismic, buoyant highs in the trench axis to interfere with subduction. There is no trench on the eastern side of the Caribbean bulge where subduction should be occurring at the juncture with the westerly-moving Atlantic plate. The Puerto Rican Trench cannot be traced any farther SE than 17°N latitude and 59.5°W longitude for a start, so we are left with the distance from there to the South American continent with no trench for subduction. There are many surface earthquakes, so we will assume that the trench has been overprinted by sediment. The trench continues to the west until it intersects the Oriente Fracture Zone, more-or-less. The earthquakes stop at about 74.5°W longitude. There is no geophysical boundary on the south.

As for events south of the Caribbean bulge, the Scotia bulge appears to be identical. Russian-German tectonic studies of the Scotia Sea region from 1994-2002 give geologic detail. Ancient Precambrian continental crust has been transformed by the process of replacement of that crust by basite-ultrabasite magmatic material and the formation of secondary oceanic crust. This is primarily by the destruction of the prior continental bridge between South America and Western Antarctica by regional stretching. The proof is of the remnants of that continental bridge still in existence in the northern and southern ridges of the Scotia Arc and in the associated Piri, Bruce, and Discovery rises in the center of that basin. The processes are listed as "manifestations of the spatial-temporal combination of processes of the continental crust destruction, thalassogenesis (taphrogenesis), and riftogenesis."

The earthquake regime for the Scotia region: 0-49 km deep has 1276 events; 50-99 km has 159 events;100-149 km deep has 138 events; 150-199 km deep has 29 events; 200-249 km deep has 1 event; and 250-299 km deep has 1 event. No deeper earthquakes exist in this region.

All of the earthquakes are essentially above the strictosphere in the surge channel/crust range. The earthquakes define a region of flow along the southern portion of this plate turning counter-clockwise to the north and northwest. By the point at 33°W longitude, earthquake activity has ceased.

The GEOSAT gives no indication of seamounts at the active margin. The earthquake regime shows a clustering of epicenters on the north and southeast portions of the active arc, suggesting the presence of a surge channel. Mapping by the use of the GEOSAT data gives the appearance of bathymetry, which it is not. It is the satellite-measured, fine-scale gravity data which shows the existence of old, cold highs and lows on the ocean floor. A map shows on the extreme northwest the Falkland Islands plateau, and on the southwest is the South Shephard and the South Orkney Islands. Proceeding through the Drake Passage on the left (west) across the southern portion of the Chile Trench, the South Georgia Ridge to the north is next. This turns south. Fractures off the Atlantic-Antarctic Ridge delineate the southeast portion. This leaves the Scotia arc as the easternmost high, and the South Sandwich Trench as the easternmost, seaward low.

Evolutionary events leading to the current geomorphology found around South America seem very comprehensive at first glance. South America and Antarctica have theoretically been separating since 84 Ma, with South America moving westward more rapidly than Antarctica. This has caused 1320 km of E--W left-lateral strike-slip displacement and 490 km of N--S divergence, resulting in the opening of the Drake Passage. An angle change during the global Eocene plate reorganization (Hawaiian-Emperor elbow, rise of the Himalayan Mountains, etc.) led to the onset of ocean floor spreading in the Scotia Sea at 30 Ma. This is presumably also the onset of subduction, creating the South Sandwich Trench during this time. Subduction retreat along the South Scotia Ridge and South Sandwich arc, combined with backarc spreading contributed to the width of the gap between the continents. The South Scotia Ridge is the trunk channel, and the Drake Passage is marked by the Shackelton Fracture Zone.

Looks can be deceiving. Preferential eastward flow of the asthenosphere due to the Earth's rotation, drag and the Coriolis force being the primary components, leaves it's clue by telltale bathymetry. Almost all of the Pacific convergent margins show an eastward bowing. R.M. Russo and P.G. Silver (1995) attribute the eastward-pointing bow in the Peru-Chile trench and the Atlantic eastward-pointing Lesser Antilles and Scotia arcs to the Nazca plate being pushed by South America. The American plate is much larger than the Nazca plate. Therefore, the Nazca plate is being bulldozed by the American plate and the South American craton. The roots of the continent have been theorized to be deeper than the asthenosphere, 450 to 650 km to be exact, by Woodhouse and Dziewonski. As the mantle pushes against the Benioff zone, the Andes Mountains react by partially rising and partially pushing back.

13. Mediterranean Sea

Much activity, as well as many tectonic scenarios, are hypothesized for the Mediterranean Sea region. This region lies between the African and Eurasian plates, containing any number of microplates, such as the Anatolian, Apulian, etc. depending upon which tectonic scenario one reads. Essentially, the African plate appears to be subducting the Eurasian plate. The location of a triple junction where the Anatolian plate encompasses the northern portion down to Cyprus across Syria, the African plate lies to the west of the Dead Sea Fault, and the Arabian plate lies to the east of that, stopping at the Bitlis Suture Zone. This gives a subduction zone comprised of, starting on the west, the Hellenic Arc running into the Cyprian Arc to the Taurus/Biltis into the Zagros Crush Zone and the Makran Margin.

However, earthquake distribution provides no clue as to the location of the subduction zone (Appendix B-1), particularly in the region of the Pliny/Strabo trench system and in the Pythias and Giarmann trenches in the west-central basin. That activity is comparatively high along the Hellenic Arc and East Anatolian Fault Zone though.

From Russian seismic surveys, the low velocity layer is of the Precambrian African platform, and that crust extends to Cyprus. Limestones on top of Eratosthenes Seamount are Upper Cretaceous. No real evidence exists for oceanic crust under the frontal zone of the Cyprian Arc, but the authors believe that the crustal structure represents tectonic activity at a convergence zone related to subduction processes. Different aspects of a deep geologic structure also exist there. Messinian evaporite layers are intermingled.

In a massive effort, John Hall of the Geologic Survey of Israel has co-authored books on the Levantine Basin and Israel in 2005. In one of the chapters, he and David Neev go into great detail about the spiraling geosutures passing through Mother Earth. The locus of their hypothesis lies in the Alpine-Himalayan orogenic belt, the driving force of which is explained by whole mantle convection processes. Their primary points are: (1) the extension of oceanic fracture zones through the continents, (2) continued tectonic activity along the fracture zones since the Precambrian, (3) functioning of whole mantle convection processes, combined with Earth rotation as the chief mechanism, and (4) Laurentia and Gondwanaland. In this hypothesis Gondwana has drifted northward since the Early Phanerozoic, that is, since the onset of the Cambrian, with no polar wander.

The Great African Plume seems to drive most of this model, which is admittedly centered on the eastern Mediterranean basin. Upwelling and downwelling exist. Plumes incorporated into the Wilson Cycle take this model back to at least the Proterozoic to Early Paleozoic, that is, about 2.5 Ga. Between the Cambrian and Cretaceous the NE African-Arabian Shield gradually subsided to be transgressed and covered by sediments. The location of this activity was from Morocco through Algeria, Libya and Egypt, Sinai/Arabia, southern Iran, to terminate at the Punjab. This all passes through a series of vortices they call the Sestri, Aegean, and the Kersihir to begin with.

The different aspects of their Mediterranean model will be discussed at the proper time. This model is merely introduced to give a background.

14. SW Asia Margin

The Makran convergent margin extends undersea from the Owen Fracture Zone on the east to the Strait of Hormuz on the west as a series of parallel ridges going onshore to the northwest to continue as the Zagros Crush Zone, eventually bounded the region known as the Dasht-e-Lut in central Iran (Figure 52). Rising to the north from the 3500-m plain, the E--W-trending ridges and valleys dominate the bathymetry. These ridges are a foldbelt, with the relief diminishing upslope on the eastern portion so as not to continue ashore. Minimum depths for the ridge axes, starting from the plain, are 2900-m, 2600-m, 1800-m, 1100-m, and 700-m superimposed on a regional slope of 1.5^{o}. The ridges range up to 1830-km long. The compression axes range from 3 to 7 km with the ridges becoming shorter

Figure 52. 3D of the Makran margin from the south where the bathymetry is married to the topography to show the effects of compression in this region. The parallel ridges go from the seafloor onshore as the Zagros Crush Zone. The Dasht-e-Lut of Iran, a circular feature called a vortex structure, is in the background. Any magma flowing out of here would necessarily feed the Himalayan system to the east.

upslope. Below 1800-m the ridges are more fully defined. Inter-ridge swales average 400-m deeper than the ridge crests throughout the entire province. This fact suggests a uniform rate of convergence. Depositional thalwegs exist in the upper swales, but there is no sedimentation between the lower ridges. V-shaped valleys are the norm in that regime.

The crush zone is characterized by a series of generally straight canyons connected to lower sloped channels. These channels take the sediment load out and create fans from 1400-m to 2200-m within the ridge province. The distinctive fan morphology loses its character as it progresses to the west, presumably because of the current outflow from the Persian Gulf. This current could also be responsible for sweeping the lower inter-ridge valleys. The deeper passage of the several of the channels passes out onto the plain. On the physiographic diagram there are three distinct, arcuate debris pediment lobes offshore that terminate before the ridge province. Within those lobes are several rotated compression ridges. Seismic data reveal that these ridges are younger and/or less active than the deeper ridges.

The core of Arabia is comprised of Archean rock. The northern boundary, the Oman and Zagros Mountains, is structurally similar, consisting of the Precambrian core covered by Cretaceous sediments. The region has been thrusted, faulted, and uplifted. In the plate tectonic regime, these mountains and the Gulf of Oman are the plate's leading edge. The northwestern portion is called the Zagros Crush Zone, and the southeastern portion is called the Makran subduction zone.

No trench exists here, and almost all of the earthquakes lie above the asthenosphere (Appendix B-1 and B-2). For now, the earthquake regime shows no deep earthquakes and only four intermediate: 0-49 km deep has 4670 events; 50-99 km deep has 859 events; 100-149 km deep has 590 events; 150-199 km deep has 453 events; 200-249 km deep has 564 events; 250-299 km deep has 46 events; 300-349 km deep has 3 events; and 350-399 km deep has 1 event. No deeper earthquakes exist in this region.

Similar crush zones appear off the coast of California (the Borderland), the coast of British Columbia and onshore in the form of the coast ranges, and off the west coast of the Philippines in the South China Sea. The Coast Ranges of California, like the history of the Appalachian Mountains, cannot be explained by plate tectonics. The same happens in British Columbia and Alaska. Features there are called exotic terranes for the want of any source of alien material. This hypothesis is unsupported by the field data. Earth contraction could be the cause of such a geomorphology.

Closing out this chapter, the trench is the primary feature of the Types 1, 2, and 3 active margins, with other features acting or interacting around the trenches. Their lengths will be important to a later discussion. They are: (1) the Pacific Ocean basin has the Kuril-Kamchatka (2200-km), Japan (800-km), Izu-Bonin (800-km), Mariana (2550-km), Nankai Trough (500-km), Ryukyu-Philippine (1400-km), New Hebrides (2050-km), Vityaz (includes variously called Kilinailau, North Solomon, Ulawan, and Cape

Johnson segments; 2750-km), Tonga (1400-km), Kermadec (1500-km), Peru-Chile (5900-km), Middle America (2800-km), and Aleutian (3700-km), (2) the Atlantic Ocean Basin has the Puerto Rican (1550-km) and the South Sandwich (1450-km), and (3) the Indian has the Andaman-Nicobar-Java-Timor (4500-km). The global linear distance of the trenches is 34,450 km. Some readers may object to the omission of the Cascade region. No trench exists there. The Aleutian and Vityaz are included, even though they are also suspect.

Where continental crust is in collision, the Type 4 active margin, no subduction occurs. Therefore, no trench occurs. In its place is usually a mountain range, such as the Himalayas. The total distance of the collision active margins is 9500 km. This gives the total take-up margin linear distance at 43,950 km, a little over half the distance of the spreading centers.

PASSIVE MARGINS and SEDIMENTARY FEATURES

The ocean floor is also home to many other feature types. Secondary tectonic forces are related to sedimentation, both erosional and depositional. These forces have a tendency to flatten, broaden, and overprint the bathymetry. Examples of sources of sedimentation include terrigenous, volcanigenic, eolian, biogenic, authigenic, and extraterrestrial. The agents of delivery to the environment include turbidity currents, volcanoes, wind, bottom currents, mass wasting, earthquakes, and in situ marine creatures. Post depositional processes include biogeochemical, physical/chemical, and mechanical means. Much evidence exists for deep ocean sedimentary features, too.

The passive margins are generally features such as continental slopes, shelves, and basins. They present their own geomorphology, which are more or less affected by these secondary processes. The slope shows a typical shelf break at 130 m, which coincides with the last eustatic sealevel fall. The upper slope at a passive margin will be incised by canyons, seachannel heads, and deltas, while the lower slopes will support fans. The borderlands and marginal seas, or backarc basins, behind island arcs are examples of these. Shelves are relatively flat and shallow near islands or continents, which from the sea's perspective are nothing more than larger islands. Seachannels generally carry most of the sediment to low areas on the ocean floor that fill in to become the basins and abyssal plains. As the sediment load is dumped at the distal end, a "dam" is built. The seachannel then makes a turn and continues until this happens again and again. Eventually the regional low is reached, and the seachannel generally dissipate there. The Maury Seachannel may be the most well known. Outer slope and basin features include fans, contourites, drifts, and abyssal plains with only a few meters of relief.

Examples of sedimentary-type features are shown with more lengthy discussions of the bathymetric characteristics, mainly in conjunction with other types of features:

1. Canyons

When enough data had been collected to notice the deep cuts on the continental shelves, the existence of canyons became known. The first one discovered was at the mouth of the Congo River by Buchanan in 1887. Passive and active continental margins are home to canyons. Canyons generally occur on continental margins of greater than 1° slope. Sometimes they are the underwater extensions of onshore rivers, such as the one off the Kamchatksky River. Canyon morphology is very similar to that of fluvial systems on land. Because they are active, they were thought to have V-shaped profiles for the most part, but this has since proven to be a function of the older sonar collectors. Canyons have U-shaped profiles also.

Canyons include both linear and meandering and both pinnate and dendritic drainage patterns. Multibeam and side scanning sonar mapping instruments and techniques are now available to go along with visual observations of submarine canyon geomorphology. These visual observation methods, such as ROVs, reveal such things as the overhanging cliffs in the Scripps Canyon, which would never show up on surface derived bathymetry. Several features are common to most submarine canyons. They have sinuous courses, their floors deepen seaward, they lose their V-shaped profiles at the same distance from shore, they are rarely found on gentle slopes, they cut through any kind of substrate, and they have tributaries.

Various tectonic processes are responsible for the formation of canyons. Canyons off the northeast United States were originally downcut by the subaerial erosion of rivers into the slope during the Pleistocene Ice Ages when the sealevel was lowered. This is also true for canyons off the Kamchatka Peninsula. Fracture control has been suggested for the formation of some western Pacific forearc canyons. Mass wasting by turbidite scouring and up- and down-current flow aid in increasing the size of the canyon. In some lower latitudes this is abetted by carbonate upbuilding for additional geomorphic character. Vertical tectonics may also aid in the upbuilding, as is the case with the Grand Canyon.

A. Kamchatka Canyons

The Kamchatka canyons empty into the Kuril Trench (Figure 29). No seachannels are associated with these features. Canyon formation control is by faults that cut through pre-Miocene basement in a cross-strike direction, meaning they are probably the result of subaerial erosion. They are being deepened by turbidity currents, and, in fact, owe their extended existence to the high amount of volcanism and earthquake activity on the Kamchatka Peninsula. These not only supply debris flow and turbidites but also initiate gravity flow for the sediment transport.

The Kamchatsky Canyon heads at the mouth of the Kamchatsky River and extends 175 km to the trench. The thalweg gradient is from 1-3° on the upper slopes, which are a dendritic pattern, until the lower portion, where they coalesce to form a 150 m deep U-shaped channel. Storozh Canyon has two upper valleys which join about 50 km out on the slope. The average thalweg gradient for Storozh is 5-7°. Tyushevsky Canyon has a dendritic pattern that coalesces and goes 45 km to a basin-plain. The upper and lower slopes average 10°, and the middle approximately 4-5°. Olga Canyon's V-shaped valley follows a fault zone for 66 km. The fault zone concept is worth noting, as it will play a major role in the discussion about the emplacement of the volcanoes on the active arc. The slightly concave channel has an average inclination of 3-6°. Kronotsky Canyon is 105 km long and is fault controlled. Zhupanovsky Canyon heads at several short coalescing valleys, the upper part following a fault zone, and it dumps it's sediment load 156 km across the slope in the trench. The upper canyon thalweg gradient averages 2-5° with an occasional 10°, and the lower averages 0.5-2°. The final canyon, Avachinsky, has a split head. The northern one was fed by the ancient Avcha River, and the other is fed by the new one. This valley is fault controlled, and it extends over 150 km into the trench. Its thalweg averages 1-3°. If, as this paper would have it, the introduction of the Obruchev Rise into the subduction zone has in part caused the canyons, that puts a time frame of Pliocene on the initial arrival of this large feature into the subduction zone. The forearc highs appear to be the remnants of "normal" outer arc high material. Because of the canyon incisions,

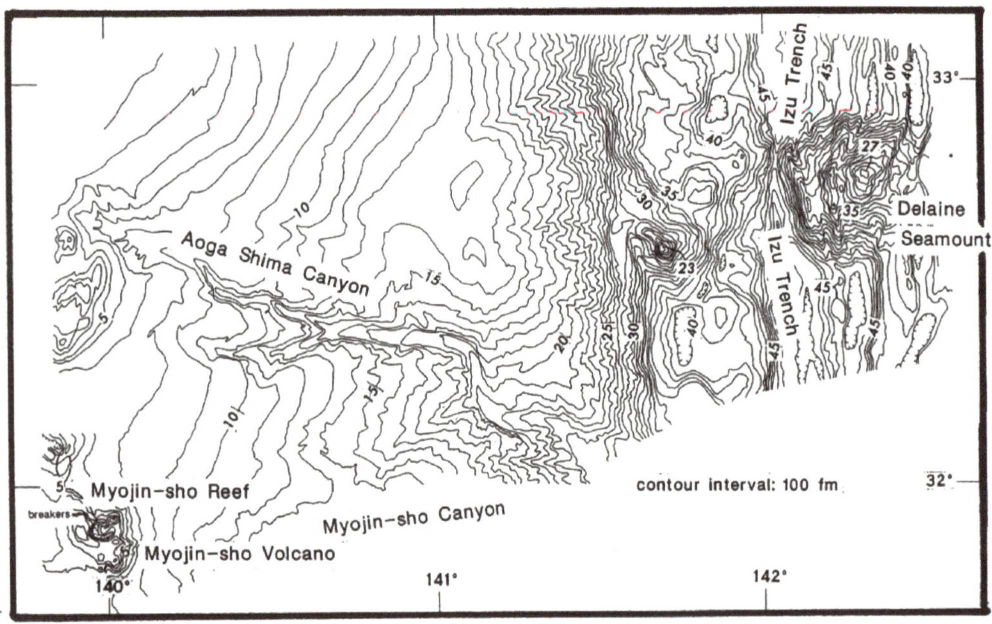

Figure 53. The Mariana forearc from the active arc to the seaward side of the Izu Trench from multibeam sonar survey data at a 100-fm contour interval. Of primary interest here is the Aoga Shima canyon/ seachannel crossing the forearc.

speculation is that these highs are not structural seamounts.

B. Izu-Bonin-Mariana Forearc

The Izu-Bonin-Mariana forearc has been pinged to death since the 1980s by academia. Before that, the Navy did the same in the 1960s. As a result, it is the most well-known piece of ocean floor real estate around (see Figure 23 for the entire region). The forearc has already been discussed with the exception of detailed descriptions of the canyons. Several canyons stand out, the Aoga Shima and the Myojin-sho Canyons.

These two 150-km long canyons were originally outlined by NAVOCEANO's multibeam surveys. The canyons were noticed to have both straight and sinuous trends with dendritic drainage patterns. They were also noticed to lose bathymetric character as they progressed further out onto the outer arc high and the upper inner trench slope. The original hypothesis was that the basin was filled by unconfined mass flows which changed to channel cutting on the outer arc high and cut headward by mass wasting. On further investigation using the side scan sonar SeaMARC II, Adam Klaus of HIG found that the upper Aoga Shima Canyon is incised into the forearc about 1200--1700-m on the 1--2° gentler upslopes (Figure 53). The conclusion is that the canyon were formed by headward erosion initiated by small-scale mass wasting which created the drainage patterns. Large-scale slumping added to the gross morphology, and the effects of mass wasting were accelerated by the addition of the volcaniclastic sediments from the active island arc volcanoes. Slumping of the canyon walls started tributary canyon formation. Channel outcrops show that most of the more recent channel outflow is limited to the channels themselves. Erosional resistant outcrops also show the possibilty of fracture control in the initial stages of canyon formation. This possibility has been speculated upon previously. The timing is suggested as beginning in the Eocene with mid-Oligocene rifting of a tholeiitic and boninitic arc massif. Both Aoga Shima and Myojin-sho Canyons skirt bathymetric highs, highs which create bends in the canyon thalwegs. Sediment erosion and rapid transport to the trench axis are the present primary intra-canyon processes.

Geomorphologic speculative interpretation from total coverage bathymetry provides a review of Hahajima Canyon's formation possibilities. Fracture control caused by intermediate to high stress in this region and high angle faulting on both sides of the Izu Trench could be caused by the arching of both the lithosphere. Under this high stress regime, the introduction of an aseismic feature, such as the Uyeda Ridge, into the region can cause tensional rifting on the inner (landward) trench wall. The overriding lithosphere can buckle if it is jammed or not moving as fast as the approaching lithosphere, which is true in this case. Enough stress from buckling can cause rifting on the overriding, landward lithosphere. In this case, Hahajima Canyon is directly across the trench from the Uyeda Ridge (the next feature to the south) and continues to the Bonin Islands. The presence of an obducted seamount on the inner trench wall lends even more credence to the stress regime (from Shoichi Oshima, Japanese Maritime Safety Agency, pers. comm.). Given the resistance to convergence by the ridge and the buckling and generally weak nature of the overriding lithosphere, this canyon probably originated as a fault caused by stress transfer by the Uyeda Ridge complex. Turbidite scouring such as is observed on the Mariana forearc could provide the mechanism for widening and deepening the fault to produce the present canyon geomorphology. One can also assume the canyon to have originated at the trench and continued to the islands if the width of the canyon is relative to its age. It is certainly wider at the trench.

C. Oceanographer Canyon Complex

The Oceanographer-Lydonia canyon complex (Figure 54) off the Georges Bank of the NE United States is primarily found on the continental slope but does not incise the shelf more than a few kilometers. The complex cannot be correlated to any landward rivers. As a result, it is not very denticulated and has only secondary bifurcations, so it was probably formed by subaerial erosion during the Pleistocene Ice Age. Studies show sand layers buried some distance from canyon heads that were probably introduced during that time. The two primary lineations, the Oceanographer thalweg and the unnamed canyon in the middle, are only about 120--130-

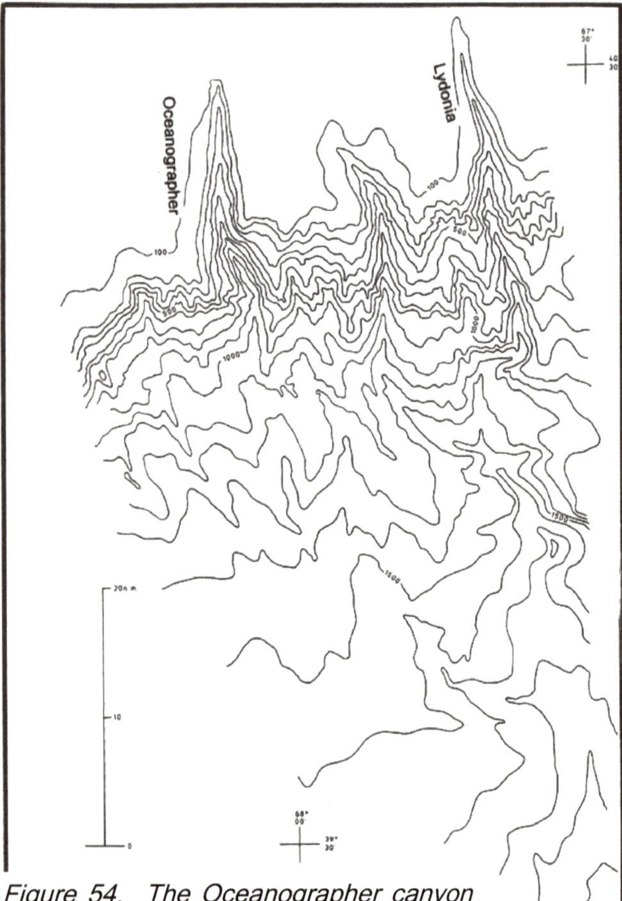

Figure 54. The Oceanographer canyon complex off the east coast of the United States is a series of four blending into one as it crosses the continental shelf. At a 100-fm contour interval, this diagram could be used as a training device to trace thalwegs.

km long. The upper thalweg slopes are 9%, and the lower slopes before the fan and Sohm Plain are about 2%, well within the outlined slopes of the definition. The canyons are also narrower on the upper slopes and deeper by several hundred fathoms. The Lydonia Canyon branches sharply west about 55 km from its headwall. The sedimentary rock there must be very resistant to erosion compared to the surrounding bedrock. Vertical tectonics of the entire Cape Cod region returning to isostacy after the retreat of the last glacial episode would have caused the initial growth of the canyons by runoff. These canyons are primary feeders for the Sohm Plain, which will be discussed further.

2. Seachannels

The next features, seachannels, serve a very real function; they are the avenue for cleaning the upper slopes of terrigenous sediments. The presence of seachannels was first detected by Hull in 1898 off the British Coast. Seachannels were originally defined as steep-walled, flat-floored, linear sedimentary depressions associated with abyssal gaps. They were thought to be 1.5 to 8 km wide and have 200--300-m of relief. Norm Cherkis, head of the ocean floor section of the USBGN (pers. comm., 1995) lends his learned opinion of the difference between seachannels and canyons. "The definitions are similar, however, seachannels are generally of much greater length, and much lower relief than expressions of submarine canyons. Further, a 0.3 degree slope is more than sufficient to keep a channel open on sea floor plains at depths of greater than 5000 meters, whereas a submarine canyon is continually being cut by riverine or estuarine efflux, normally creating slopes of 5 degrees or better. Submarine canyons usually terminate and create fans or cones shortly after the slope-gradient approaches 0.5 degrees."

Seachannels are shaped by gravity flows such as mass wasting, turbidity currents, thermohaline currents, and tectonic processes, making them the result of secondary processes. They may be fed by pelagic sedimentation, turbidites, or water flow. Many seachannels originally formed during the regressions brought on by the Ice Ages when the sealevel dropped 120 m below the present day and canyons formed. About 8000 years ago, the sealevel started to rise, locking seachannel imprints on the shelves. Many present seachannels are extensions of slope canyons that empty onto and dissect abyssal plains. Seachannels follow the path of least resistance, so they can meander, braid, and even follow fracture zones.

The characteristic bends in the seachannel axes are caused by deposition. In general, as seachannels prograde, they build a fan at the end. Eventually the fan builds high enough to prevent the passage of the turbidity currents, so that the seachannel essentially dams itself. The next large event (i.e., one with enough force and/or enough sediment load) will break through the dam or flow to one side. This is always downslope. As the seachannel continues to flow intermittently and age, this process will repeat itself countless times as it has on each of the examples included. As the fan progrades, an abyssal plain is built. Where deposition is not occurring, the seachannel axis may not meander. If it is being scoured by the passage of a relatively fast-moving body of water, and there are no changes to that flow, then the axis may be straight.

Many seachannels exist. The primary seachannels in the North Atlantic are the Northwest Atlantic Mid-Ocean Canyon at 4000-km long, the Imarssuak Seachannel at 1000-km long before braiding, and the Mid-Ocean Channel No. 2 at 575 km, all of which helped level the Sohm Plain. The Maury Seachannel was the avenue of transport for the sediment which helped build the West European Basin. The Theta Gap channels turbidity flows from the Bay of Biscay onto the Iberian Plain. In the North Pacific the Gulf of Alaska is the site of the majority of the seachannels. The Aleutian Plain is in part built by sediments of the Seamap, Sagittarius, Aquarius, and Taurus seachannels. The Surveyor Seachannel and distributaries are feeders and incisers of the Alaskan Plain. Baker, Horizon, Mukluk, Moresby, Scott, Baranof, Heck, and Tsimshian Seachannels are feeding sediments to the Tufts Plain. Further south lie the Astoria, Monterey, Cascadia, Willapa, and Vidal Seachannels. A channel on the eastern side of the Emperor Chain empties into the Jingu Basin. The Chuo-gato, Higashi-goto, Nakwe, Nishi-goto, and Toyama seachannels are in the NW Pacific Sea of Japan. The Mekong Seachannel and the seachannels and canyons coming from between Macau and Hong Kong are responsible for carrying the load to build the South China Sea Plain Bathymetry in the southern oceans does not allow one to outline channels there with the exception of the Pernambuco Seachannel off Brazil.

The Toyama Seachannel heads in the Churu-Sangaki drainage basin on Honshu Island, Japan, and is fed by at least seven rivers. The seachannel goes north from 37°N latitude, passes around the Yamato Ridge, and finally starts to deposit its load at about 41°N latitude where it fans onto the Japan Plain in the Sea of Japan. The Toyama Seachannel is at least 460-km long. The Japan Plain is of some interest to the overall scope of this treatise. Much research gives rock ages: Korea Plateau, 1.98-2.72 Ga gneisses; Ulleung Plateau, 102-110 Ma granitoids; Kita-Yamato Bank, 197 Ma granodiorite; Yamato Ridge, 178-310 Ma granitoids; and Kita-Oki Bank, 141.6 Ma granite. These ages have led Yoon to conclude that these features were emplaced in the formation of the Japan Sea, and that suggests that the Yamato Basin is a rift formed by diapiric tectogenesis along the Sado-Pohang belt.

The Gulf of Alaska has many seachannels (Figure 55). Some of the Gulf of Alaska seachannels are not contiguous with submarine canyons and are not linked with river drainage systems, the first being the Surveyor Seachannel on the extreme north. The Tsimshian Seachannel in the Gulf of Alaska arises off the Dixon Entrance by the Queen Charlotte Islands and rather circuitously winds its way south for several hundred miles to pass between the Schoppe and Peters ridges and fan onto the Tufts Plain. The stacked profiles in close proximity to each other show V-shaped, U-shaped, and partly filled channels with levees. The existence of traces of several other seachannels in the Gulf of Alaska is now verified, such the Tlingit and Haida Seachannels. The quality of the bathymetry and its density are the primary considerations for contouring these features because there are so many of them, and they wander all over the place.

Seachannels in the Gulf of Alaska mostly head at the shelf break. They are not associated with on-shore drainage basins or submarine canyons such as the other seachannels because the prominent offshore right-slip faults appear to have offset them. They are, however, associated with the Gulf of Alaska abyssal plains, the Tufts and the Alaskan, and are presumed to have built the plains and levees while continuing to do so. DSDP Sites 178-182 were adjacent to the Queen Charlotte Islands during the Miocene, about 20 Ma. Terrigenous sediment was deposited on the plain at that time by the Surveyor Seachannel, showing net movement north because of the seachannel's present location. This material was primarily basalt because of mountain uplift. About 5 Ma the prodigious volcanic activity started. During the Pleistocene, the great Ice Ages occurred, and much glacial erosion of the mountains produced angular quartz.

The fans and plains in the Gulf of Alaska provide various samples to help ascertain their origin. Site 178 is overlain by terrigenous sediments for the Pleistocene, clastics for

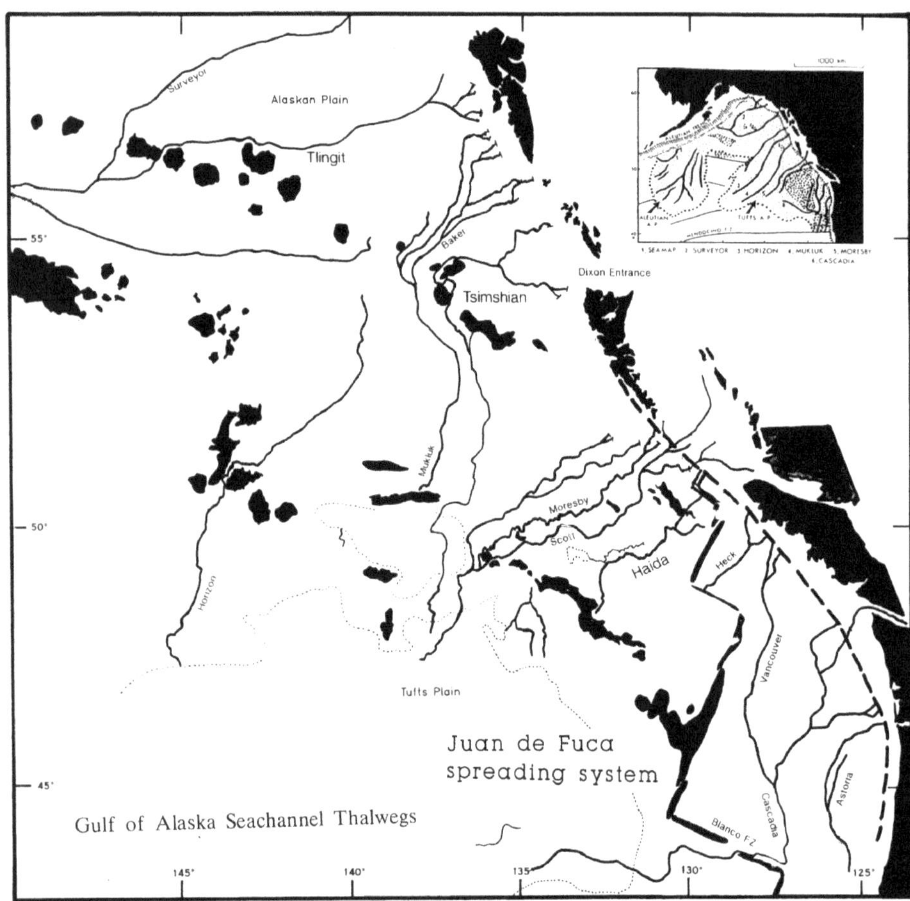

Figure 55. Seachannel thalwegs in the Gulf of Alaska derived from NAVOCEANO's multibeam sonar data base. Many were already known; many have had to have course corrections applied, as they were drawn from incomplete data sources and not delineated as they actually are. (Several named after the Indian tribes inhabiting each region)

the Pliocene, and mixed terrigenous and diatomaceous sediments before that. Site 177 shows no shallow water origins for the sediments. Site 173 is the same as 178, as is offshore Vancouver Island. Piston core, 18-8, located SW of DSDP Site 177 is somewhat south of Moresby Seachannel but on the Tufts Plain. It retrieved 95% angular quartz. The angular quartz suggests glacial processes. Any cores now on the Tufts Plain should yield rounded quartz from riverine sources. NAVOCEANO had cores out to 80 km showing mostly rounded quartz and mica (E.V. Kelly, NAVOCEANO, pers. comm., 1992). With all the tectonic forces acting in the headwall area of Scott and Moresby, this is to be expected.

No pelagic or hemipelagic sediment appear in any of the cores, proving the activity of the seachannels. The terrigenous sediments overwhelm the presence of pelagic sediments. Farther south, off the coast of Washington and Oregon at the Quinault Canyon, the sedimentary regime has been noted to have changed from turbidite to hemipelagic deposition since the onset of the Holocene transgression. Scott, Moresby, and Haida seachannels are feeding on all the material on the outer shelf by pirating. They also must have formed by turbidite scour, and that is the reason they do not join with onshore rivers. Apparently, during the tectonic uplifting of the Pliocene, the glaciers of the Great Ice Age and riverine systems, such as the Fraser River in British Columbia, have loaded the continental shelf with terrigenous sediment. Wave action during the low stand carried the sediment out to the shelf break during the Pleistocene. Tectonic activity causes turbidity currents, and these follow prescribed paths, or seachannels, to their ultimate destination either as levees or the abyssal plain. Most of the sites drilled on DSDP Leg 18 verify this scenario. That is the reason for the predominance of angular quartz on the plains. Most of the rounded quartz is reported at or near the Fraser River delta and is just starting to find its way out onto the shelf. As to the problem of headwall activity, the present high stand of the sea during this transgression has all but halted major backcutting into the slope. Riverine energy is mostly dissipated before it gets out to the shelf break to help with the backcutting. The offshore dams have withheld most of the rounded quartz riverine sediments.

A comment about previous seachannel studies is warranted before continuing. While the proper geomorphology was known, how the seachannel thalwegs were connected was not known. The head of seachannel "A" may be connected to the body of seachannel "B" may be reconnected to seachannel "A" or to seachannel "C's" fan. The newer multibeam sonar surveys available for the Gulf of Alaska allow the bathymetry to be updated. Understanding that seachannel beds are very ephemeral from a bathymetric standpoint helps with a new interpretation, especially when compared to different contour interpretations from different data bases which show different axes, or thalwegs.

Braiding is caused by gradient steepness, large sediment loads, non-cohesive bed material, widely fluctuating discharges, or rapid aggradation. This is what one would expect from an undersea "river," the same morphology that we see on land.

3. Fans

Side-scan imagery of the Bear Island Trough Mouth Fan in the Arctic Ocean shows many sedimentary features. Ice scour marks are the thin, generally NW-striking ridges on the eastern side. The northern braids are the smaller channels incising the previous deposits. The light WNW-trending lineations are the latest sediment deposit tongues. The lower flanks of Greenland are on the west. Eight seismic units have been identified as being comprised of Middle and Late Pleistocene sediments. These units were deposited during the eight glacial advances in this region. Some mounding is probably the result of submarine debris slopes, those slopes finally becoming steep enough for the contained material to overcome the angle of repose. During interstadials and interglacials the sedimentation was reduced.

4. Abyssal Plains

The last, and certainly most extensive, midocean feature is the abyssal plain. While not a subject to elicit a high degree of excitement, these features nevertheless are a large result of secondary sedimentary processes. Lying at the distal ends of canyons and seachannels, the plains are the smoothest regions to be found. The smoothness is ameliorated by pelagic and hemipelagic sedimentation. Their collective slopes are no more than 1 m/km, occurring between 3000 and 6000-m deep. Seismic surveys show that these flat regions cover the original landforms. Riverine erosion, or terrigenous sediment, is the primary constituent of these great sediment traps. Some plains do not lie at the ends of canyons and seachannels, so they do not receive the abundance of terrigenous sediments. Some active seachannels continue to incise the older plain and accrete further out on the plain, thus building the plain even further.

Plains were discovered on a 1947 Mid-Atlantic Ridge survey. The larger abyssal plains are in the Atlantic Ocean basin lie the Sohm, Hatteras, Nares, Pernambuco, Argentine, Cape, Angola, Iberia, Tagus, Madeira, Biscay, and Cape Verde plains; all deep basin areas. The Sigsbee Plain is in the Gulf of Mexico. The Aleutian, Alaskan, and Tufts plains are in the Gulf of Alaska. Otherwise. The deeper ocean floor has more of a sediment cover.

The Sohm Plain on the NW Atlantic ocean basin presents an interesting case study. In the 1960s the abyssal plains were felt to be totally flat. Lavern Snodgrass of the World Chart Group at NAVOCEANO has had to update the Sohm map three times because "all these seamounts kept showing up." The coverage of the underlying topography is not quite complete. The tops of the seamounts and ridges that are a part of the trans-Atlantic fracture zones marks the continuation of those features. Otherwise, the many near-shore canyons provide the feeder system for sediments to build the Sohm Plain.

With the various rivers emptying terrigenous sediments from all around the Mediterranean basin, the plain surrounding Eratosthenes Guyot (Figure 56), and including the guyot itself, should show all of the predominant secondary and tertiary forces acting on any part of the ocean floor. And, it does. Lying between Cyprus on the north, Lebanon and Israel on the east, and Egypt on the south, this is a case study in depositional environments. To begin with, vertical tectonics and the low stand exposed this section of the Mediterranean to subaerial erosional forces in the not too distant past. This allowed rivers such as the Nile to continue out onto the shelf even further to dump their loads while also eroding Eratosthenes Island itself. The results of that are evident in the figure.

Eratosthenes Guyot is located at the extreme eastern edge of the Mediterranean at 33°50'N latitude and 32°40'E longitude. Some form of tectonism had rent Eratosthenes into at least three recognizable segments after the initial erosion and mass wasting required to plane the island into it's rough guyot morphology. The azimuth of WSW-ENE may provide a clue as to those tectonic events. Recent studies do nothing to clarify matters. "It remains unclear what type of crust underlies Eratosthenes Seamount." The multi-channel seismic data are fraught with question marks at depth. It has been determined, however, that Eratosthenes is a continental fragment.

Additionally, there seems to be some evidence of compression between Cyprus and Turkey in the form of beach-parallel, longitudinal ridges similar to those of the Zagros Crush Zone and the Makran convergent margin. This would mean that this section of the Mediterranean basin is undergoing crustal shortening, just as is the Arabian Sea basin. That is to be expected in a constantly contracting Earth.

After those events, a high stand of the sea ensued, and active submarine erosion continued, and continues, to change the regional bathymetry. The initial sediment deposits are now incised by younger seachannels, which are dumping their loads further out beyond the shelf in the basin. On the northwestern portion of the basin floor, sand waves in a whorl pattern are evident. In a region of so many vortex structures, such as the Aegean Sea, Adriatic Sea, and Dasht-e-Lut, it is entirely conceivable that this is yet another vortex.

Figure 56. The eastern Mediterranean basin from multibeam sonar data bases showing the erosional patterns associated with the shelves and guyot in that region. Eratosthenes appears to have been bisected by fractures while sitting atop continental crust.

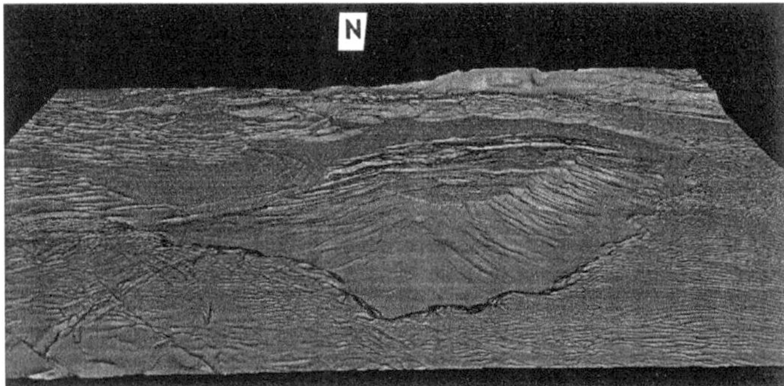

Eastern Mediterranean basin

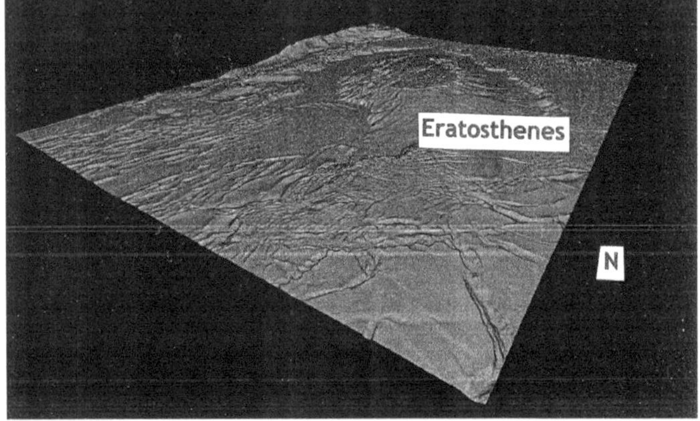

The South China Sea, located between the Philippines on the east, Southeast Asia on the north and west, and Indonesia on the south, has several explanations as to its formation and purpose (Figure 57). This 2,974,600 km^2 basin has been suggested to be the result of both indenter and extrusion processes, both involving the emplacement of India. The indenter part of the equation calls for India to crash into the Eurasian continent. The extrusion model calls for the South China Sea to have formed by seafloor spreading between 45--17 Ma, and that opening to have stopped between 20--17 Ma when the major uplift of the Qinghai-Tibet Plateau began. This allowed room for the Indochina and China blocks to extrude to the east into the open space provided by the basin. In another model, the intrusion results in crustal shortening and underplating, but this does not affect the South China Sea basin.

Much work has been done on this basin, and all of it refutes the above. The extrusion process would have magnetic lineations aging towards the SE. They do not; they age towards the NW. The time of uplift of the Qinghai-Tibet Plateau is placed at 5 Ma. The underplating calls for crustal shortening of 800--200 km. That figure is closer to 75 km. Last, the craton roots are predicted to be on the order of 650 km deep.

The South China Sea basin is there because of a rapid subsidence during the Villfranchian Era of 1000- m (Department of Marine Tectonic Geology, Institute of the South China Sea, 1988). Before that, during the Proterozoic is was at base level, just as were the Talemic, Eastern Mediterranenan, and Adriatic platforms. In fact, the South China Sea is suggested to have been a peneplain of the central Late Precambrian platform.

In the old geology there were platforms and geosynclines. The platforms were the broad, stable provinces which consisted mostly of igneous and metamorphic rocks in undeformed horizontal layers. The

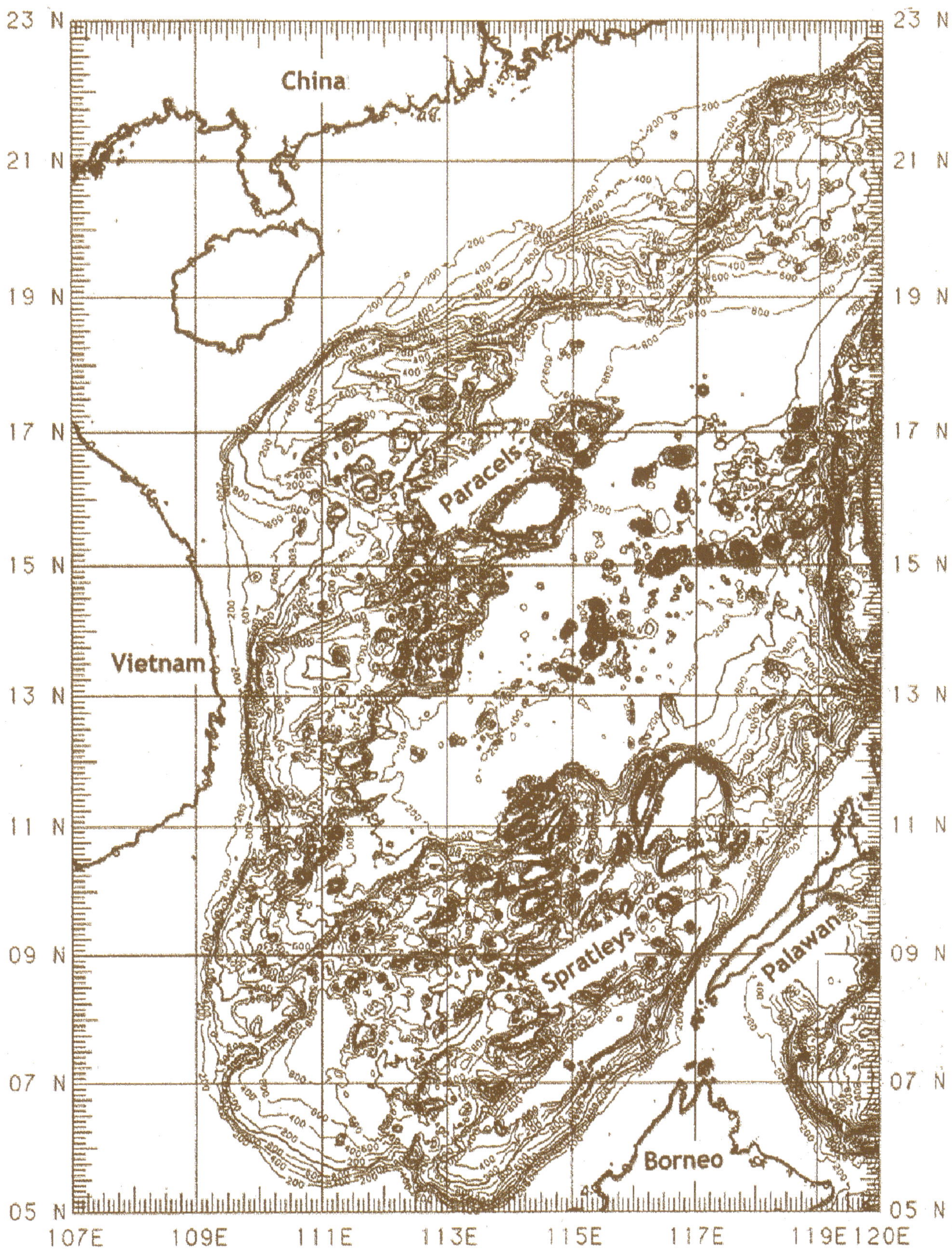

Figure 57. 1998 bathymetry of the South China Sea from multibeam sonar surveys at a 200-m contour interval on a Mercator projection. The NE-trending features lead one to believe in control by fluid mechanics from two separate sources. The basin has been in place for the past 1.7Ga.

geosynclines were the mobile belts which are covered with sedimentary layers of rock subject to movement, or tectonics. The geosynclines were built into mountains. During the Late Precambrian, about 1.7 Ga, the platforms were subjected to uplift, intensely denuded, and they became peneplains. The peneplain subsided and became the base depth for the basin. In this case, that is at the 11 km depth.

Assuming the 11 km depth to basement to be correct, the sediment load on the plain is now 6800 m because the bathymetric base depth is 4200 m. The overall strike of the plain is NE with a seamount chain bisecting the plain on the same strike. The southern portion of the South China Sea is covered with seamounts in a random pattern, including the Spratley Islands. The northwestern portion displays the same configuration of seamounts, and it includes the Paracel Islands. The eastern portion, abutting the Philippine Islands, is characterized by a series of ridges, presumably compression formed and reflective of the California Borderland and the Makran Convergent Margin. While it has been speculated that a trench lies in this region, the bathymetry shows no sign of a trench.

At last we come to the end of the feature descriptions. Given the veritable plethora of information collected in the last 40 years, much of it not being in agreement with the current working hypothesis, the time is ripe for a tectonic paradigm shift. The features and timing are in place.

TECTONICS

Most feature-types for the world's ocean basins have now been introduced through the various surveying techniques. Many of them have been age dated. The active margins have been shown in depth by means of earthquake seismology and seismic stratigraphy. The time has come to try to explain how the bathymetry and geochronology are used to help explain the tectonics to deliver the geomorphology of each feature-type.

To begin with, several different types of tectonic hypotheses are currently being bandied about, the plate tectonic hypothesis being that which is mostly in vogue. Expansion tectonics is garnering a goodly following in Australia and Europe. Several investigators are looking at wrench tectonics, where rhomboid features rule the day. Oceanization has been around for a while, as well as geosynclines. Surge tectonics offers explanations for heretofore unexplained tectonic phenomena. And so on...

1. Geometry

Before getting into the tectonics of the ocean basins, a problem with geometry may have been the impetus for new earth evolution explanations; simply put, the math does not work. Even without decent bathymetry of the ocean floor, the midocean ridges, trenches, and collision margins have been known since the mid-1960s. The ridges produce a discrete amount of ocean floor, so the collision margins must necessarily be the avenue of destruction of that same amount of ocean floor. The midocean ridges measure about 75,000 km (Figure 3), and this means that about 150,00 linear km of new material is constantly being produced according to the current hypothesis. In theory, that much linear distance in collision margins must exist to keep Earth from having a middle-age spread leading to another "Big Bang" situation. There is not; there are only 35,850 km of trenches, and that figure includes the highly suspect Vityaz and Aeutian "trenches." Including the Mediterranean-Zagros-Himalayan-Indonesian collision zone to the Timor Trench only adds 9500 km more to the take-up figure, certainly much less than the amount being produced at the spreading ridges. Surely this information leads one to suspect that Earth is expanding.

Geometric problems abound. A plate has to have a spreading center and a convergent margin. The realization that the Antarctic plate has no convergent margins whatsoever should have raised a red flag. The African plate (Figure 12) is surrounded on the west, south, and east by spreading ridges. The only thing that may be construed as a convergent margin is the Aegean Trench, a rather minuscule feature between Africa and Greece. Similarly, no convergent margin exists for the proposed Eurasian plate (Figure 3). On a smaller scale, no convergent margin can be found for the Gorda, Juan de Fuca (Figure 52), and Cocos (Figure 38) plates, although much taxpayer dollars are spent trying to locate a trench at the Cascadia margin off Washington and British Columbia. It does not exist in the bathymetry; it does not exist in the earthquakes (Appendix B-6). Time and money are being wasted looking there. The Manihiki and Magellan plates are nothing more than large igneous provinces; rises as it were. They do not have either a spreading center or a convergent margin. Lastly, no plate boundary exists between the North and South American plates or the Caribbean and South American plates. The plate adherents can take heart, though, the Pacific, Arabian, and Philippine plates all fit the classic definition for the most part.

2. **Gondwanaland** (with Ismail Bhat)

Gondwanaland is an integral part of most tectonic hypotheses. It was the southern super-continent until about 200 Ma. This super-continent was composed of South America, Africa, Madagascar, India, Australia, and Antarctica.

Eduard Suess, a Swiss geologist, introduced the concept of a Tethyan Sea in 1893, and Celal Sengor carried his idea one step further in 1984 by postulating two Tethys Seas.

An updated variant of the Neo-Tethys model, which seems to have found a broad acceptance by the cognoscenti, envisages the origin of this ocean, during the Permian, along the southern margin of a basin that had existed since the Precambrian on the northern margin of Gondwanaland; that is, along the northern boundary of the Indian subcontinent. Remember, India was supposed to be way down south joined to Australia, Antarctica, Africa, and South America. While this model had evolved into a radical shift to that of the original premise, proponents of this model neither support nor refute Sengor's model, nor do they specify the relationship of this Precambrian-Permian basin with his Late Paleozoic-Mesozoic Neo-Tethys or Paleo-Tethys Seas. It also leaves unaddressed the underlying problem that Sengor's original model was supposed to have solved.

In another explanation, in 1992 Shankar Chatterjee theorized India to have evolved when, in the Early Jurassic, the Indo-Seychelles block and Africa were attached. Subduction started between 90 and 110 Ma, that convergence being defined by the Ornach-Chaman Fault on the west and the Shan Boundary on the east. In the Late Cretaceous the Indo-Seychelles block had pivoted and was proximate to Somalia and Arabia, bounded on the north by a narrowing Tethys Sea. At the Cretaceous-Tertiary (K-T; 65 Ma) boundary India and the Seychelles separated, with India moving to the north. Complete suturing between India and Asia took place about 50 to 55 Ma, this age also being placed at 45 Ma.

All of this earlier movement should have created the sutures found in the present-day continental land mass. One of plate-tectonics' problems is that the model does not recognize that all tectonic belts have a common origin, and that compression, strike-slip, and extension all take place at the same time in the same province over the same surge channel. Now we find that the massive collision never took place. Were the India/Asia collision to be a fact, a continent/continent convergence zone would exist, and that is called the Indus-Yarlung Zangbo Suture Zone by the plate tectonicists.

Sutures, or old collision zones, have been theorized where belts of ultramafic rocks exist on continental cratons. Increasingly, detailed field work has shown that these are not zones of collision at all. Many of the proposed sutures in Asia are now recognized as old tensile, rifting, tectonic environments where mafic and ultramafic rocks have risen to the surface along fault systems. This means that the theories about Asia being a collage of micro-continents is in serious doubt.

Truly, our understanding of the evolution of the Indian region is in a quandary based on all of the previous information listed herein. That is because of the constraints emplaced by a hypothesis that was formulated before any real data were collected, a hypothesis based on sliding plates around manually to see where they fit best. The paleobiogeographic evidence points to an earlier collision or no movement at all with the exception of vertical tectonics. The actual rock samples also point to an older collision. The magnetic anomaly data are sketchy at best to support a more recent collision.

Which is best explained by the data? The entire story of what has happened at the Indian-Eurasian interface has been misrepresented long enough. The Wadia Institute of Himalayan Geology, a non-profit organization whose sole purpose is to establish exactly what has gone on there, has sponsored many papers through the past 20 years that should have been used by all concerned as a basis for any hypotheses about that region. The formation of the Tethys Ocean from the Late Carboniferous to the end of the Cretaceous involved lithosphere uplift, stretching, and the development of anastomosing normal faults. This covered the region from Kashmir to Kempa Dzong. The well-preserved continental margin suggested a maximum Tethys Ocean width of 500 km. Unexplained at that time was the age of the Ladakh batholith, an age of 235 +or- 13 Ma, the occurrence of radiolarian chert within undeformed beds, and the suggested involvement of Earth's rotation in any driving mechanisms. Next proved that pre-orogenic rift tectonics were of critical importance for modeling the structural development of the Himalayas. A long history of spasmodic rift tectonics was shown to explain actual events in the Himalayan region while showing that the Indus-Yarlung Zangbo ophiolite and other geological data represent mafic-ultramafic diapirism of an extremely attenuated lithosphere. The synchroneity of the rifting spasms is in close accord with the changes in spin-rate of the Earth.

None of the previous models addresses the stratigraphic and tectonic records of one of the best-known segments of the Neo-Tethys-the Himalayan belt- where the southern passive margin of this ocean preserves a concordant stratigraphy from the Late Archean to the Eocene, a rock sequence recoding events for the past 2.51 billion years. And, this concordant record has been known for at least the past 20 years, in English, and published ubiquitously.

That is right. Ismail Bhat and Ashok Dubey, veterans of the Wadia Institute of Himalayan Geology, have collected rocks all over the Himalayan region, analyzed their composition and age, and repeatedly published the results:

Following the initiation of basin formation in the Himalayan region during the Late Archean rifting event and associated mafic volcanism (Rampur flood basalts), the sedimentation history reveals a progressive northward shift in the fall line and depocenter that was determined by a similar shift in the locus of the four repetitive rift-related mantle-derived magmatic phases over the time until the lithosphere rupture was completed during the Late Cretaceous. Unlike other rifting phases, the early Paleozoic rifting was not associated with mafic magmatism. Instead, it caused large-scale crustal anatexis which produced a long, linear belt of leucocratic granites in the Higher Himalayas. All the rifting events are marked by the uplift of the basin, the basin which eventually became the Himalayan Mountains we see today. While the Late Archean and Early Paleozoic rifting phases resulted in the establishment of stable rift shoulders, such as the Shield and Lesser Himalayas respectively, the later events failed to do so. This is typically seen in the collapse of the stable rift shoulder following the Late Paleozoic rifting. The difference in post-rift sedimentation could be attributed to the failure in the establishment of a coupling between the surface processes and tectonics.

The Middle Proterozoic saw the Bafliaz and possibly the Abor volcanics, the Late Paleozoic-Early Mesozoic the formation of the Punjal Traps, and the Early Cretaceous was witness to the creation of the Indus-Yarling Zangbo suture zone ophiolites. These last features, the IYZSZ ophiolite, are attributed to the Tethys Sea depositional regime. However, the lithographic section of this ophiolite, with an interbedded succession of sedimentary and mafic/ultramafic litho-units and a sill complex, dispels any myths concerning the Tethys as an open ocean. It has neither an interbedded succession of sedimentary and mafic/ultramafic litho-units nor a sill complex, both of which are part and parcel of the plate tectonic definition of an ocean floor ophiolite complex. The hard-core field data instead conform to a high-density magma upwelling in a basin having undergone severe extension and rapid sedimentation.

The high-mountain marine fossils of the 26 km thick concordant sediment record, pre-Carboniferous marine fossils are found in the first 16 km. This sediment pile lies just south of the IYZSZ, which happens to coincide with the proposed subduction zone of the Neo-Tehys Sea. Sengor has no room for a seaway during this time, yet the depositional record contains over 600 Ma of these fossils. Late Proterozoic-Cambrian marine fossils are overlain by huge stromatolite-bearing carbonate deposits all along the Lesser Himalayas. Additionally, the oldest Lesser Himalayan lithologic units are co-eval to those of the Aravalli-Vindhyan region of the Indian shield.

Previous theories had all relied on internal folding and stacking in the upper crust and underplating by the Indian-Australian plate lithosphere. The geometry around the Simle Klippe region shows the opposite to be true; showing instead that folding and thrusting developed simultaneously. Reactivation of pre-existing faults puts severe limitations on the amount of crustal shortening, and, in turn, crustal thickening. An independent study showed the simultaneous development of noncylidrical folds, frontal ramps, and transfer faults in this compression regime. Last, mafic volcanics in the Kumuan Lesser Himalayas have yielded ages of 2.51 +or- 0.08 Ga, ages which are the same as for the Rampur-Mandi volcanics. Rift-related volcanics are also known to be of the same age in the Arvalli region of the NW Indian Shield. The orientation of the rifts in these regions suggests the possibility of a mantle plume setting for this region at that time. And, what this shows is that India has been in place for at least the past 2.51 Ga and never a part of "Gondwanaland."

The earthquakes offer no support for either downbuckling or underplating because there are essentially none below 250 km: 0-49 km deep has 3450 events; 50-99 km deep has 811 events; 100-149 km deep has 692 events; 150-199 km deep has 444 events; 200-249 km deep has 557 events; 250-299 km deep has 47 events; 300-349 km deep has 2 events; and 350-399 km deep has 1 event. No deeper earthquakes exist in this region.

In summary, the Himalayan stratigraphic record following the Late Archean basin formation suggests that sedimentation occurred from the shield region until the Early Cambrian all the way up to the Higher Himalayas. Marine conditions came to an end by the Early Paleozoic for both the shield and Lesser Himalayan regions. It continued to the north where the basin underwent differential uplift and erosion with widespread granite magmatism, accompanied by sedimentation. This produced the Higher Himalayan Cambrian-Ordovician granites. From that time forward not any of the Himalayan region was under water, so India did not "dock" where it is 15 Ma. There was no Neo-Tethys; there was no Gondwanaland.

Assume for a minute that we the studies of the Wadia Institute of Himalayan Geology were unknown; assume that India drifted across the Indian basin for about 50 Ma.

Remember little **Lystrosaurus**, the pig-like critter used to prove the continental drift of India. India, and tiny **Lystrosaurus**, are integral parts of Gondwanaland. **Lystrosaurus** has been found north of the Taurus-Zagros-Indus-Yarlung suture zone; that is, in Asia proper. A study mapping all of the **Lystrosaurus** remains included Antarctica, Africa, the Moscow basin in Russia, India, western China, and Vietnam. The sample base increased, but to date none have been found in the Americas. In fact, four amphibian families have been found in the same bed only in Antarctica, South Africa, and Tasmania. All four are found from as far away as Asia and Svalbard, but nowhere else are they in the same bed. Speculation now has it that **Lystrosaurus** evolved in the north and migrated south during the Triassic. Ergo, **Lystrosaurus** cannot be used to prove anything about drifting, and India has been found to be close to Asia and even Africa, but never close to the Gondwanaland continents.

Art Meyerhoff, his daughter, Donna, Mac Dickins, and Art Boucot wrote a treatise that was published by the Geological Society of America as its Memoir 189 in 1996. The biogeographical data presented in the study establishes, beyond any measure of doubt, the impossibility of any plate model to account for the distribution of 'Phanerozoic faunal and floral realms of the Earth'. Instead, in complete contradiction to the concept of Gondwanaland and Laurasia, the study establishes a broad, up to 5,000 km wide, east-west intercalary zone, "extending from western South America to the Pacific shores of Asia, Australia and New Zealand", where northern and southern fauna and flora are mixed.

Australia is another piece of Gondwanaland. J. MacGregor Dickins made a life study of the region around Australia, herein summarized: "Seismic section...western and southern Australia...formed by very large high angle tensional faulting structures generally of Jurassic age but not older and apparently not younger than Early Cretaceous...no listric faulting on the adjacent slope and continental margin. Unpublished data from the opposing margins of Antarctica show the same characteristics. The simplest explanation is that we are seeing foundering of the oceans thus forming the continental margins. There is no evidence of lateral movement of continents to justify the so called breakup of Gondwanaland..."

Further dating this region, during the Triassic Australia was largely isolated. **Kannemeyeria** (**Cynognathus**) Zone fossils are only in Australia during the Early-Middle Triassic on through the Cenozoic. Antarctica was isolated as well. Another real proof lies in the fact that many dredge hauls on the Australia-Antarctic Discontinuity (supposed spreading ridge between the two continents) yielded nothing but granites. Both Australia and India are part of the fabled Gondwanaland. Antarctica has not been in close proximity to anywhere except for southern South America.

Crocodile bones have been found in both Antarctica and Greenland. As crocs usually frequent the warmer climes, this poses a problem to the non-drifters. However, a mere tilting of Earth's axis of something like 23° will get the polar regions warm enough for **Glossopteris** plants and crocs to live. This much tilt has already been proven by NASA, as the pull of the Moon provides this agent. Based on the lineament trends of the megatrends, this angle of tilt has varied as much as 120° in the past. So, the continents did not have to display any degree of upward mobility.

Choi, in a synopsis of rock samples from the South Australian Plain, western offshore Tasmania and the Tasman Rise, and the Dampier Ridge, has found that no DSDP drillings have ever reached basement under the Jurassic and Cretaceous basalts. Dredges from the Southern Ocean and the Tasman Sea yield Precambrian granodiorites, Early Paleozoic grey-green and black shales, Early Paleozoic to Proterozoic fine-grained meta-quartzite, Precambrian basement metamorphic and volcanic rocks, Mid-Permian to Proterozoic granite, gabbro, and felspathic sandstone, and Precambrian to Early Paleozoic gneiss, granites, etc.

In the book **Geologic Structure and Origin of the Pacific Ocean** (2005) Vasil'yev, Choi, and I.B. Mishkina write that the "oceans around Australia were formed by subsiding continental blocks without essential horizontal movement. The process started in the Late Triassic in the west of Australia, and in the east and south in the Late Cretaceous and Paleogene. The area that is presently occupied by deep trench in the south of Australia had been land until Late Cretaceous which supplied sediments to the Otway Basin.""

In the same book, Udintsev and N.A. Kurentsova show that the ancient Precambrian continental crust which lay between South America and Antarctica has been getting stretched since long before the proposed breakup of Gondwanaland.

The Naturaliste Plateau is one of several large submarine plateaus extending from Western Australia into the Indian Ocean basin. The Mentelle Basin and Naturaliste Trough lie between the mainland and this plateau. The plateau is bordered by the Perth Abyssal Plain in the northwest and the Australian-Antarctic Basin in the south. Covering an area of 90,000 km^2, the shoal depth is 1550-m and the basin is 5000-m deep. In 1998 this plateau was found to consist of continental high-grade gneisses.

This still leaves South America and Africa, that is, the South Atlantic Ocean basin. While this region has not been as heavily sampled as other spots in the world's ocean basins, it has at least been sailed over several times by collectors. As noted, the oil companies are willing to show four stratigraphic units for the Walvis Ridge and it's onshore counter-part, Namibia. The pre-rift Karoo section contains preserved Permo-Carboniferous-Triassic age samples, and this section continues offshore, with the Walvis Ridge being implanted at a later date. But, that can only take place south of the equatorial region, as has been repeatedly proven. Additionally, an attempt to match the continent-ocean boundaries of South America and Africa along the salt basins has also failed. According to Ian Norton, this means that the margins are not isochronous between South America and Africa.

India, Australia, and Antarctica were never joined. South America and Antarctica had a land bridge during the Precambrian. Therefore, Gondwanaland never existed. South American and Africa must still be dealt with but, in light of the above discoveries, the same scenario will probably hold true. With the removal of the mythical Gondwanaland from any meaningful discussion/hypothesis, the data are now allowed to speak for themselves.

3. Geochronology

Having dispensed with the obvious geometry problem and Gondwanaland, the next basin-wide item is the rock ages, the geochronology. Certainly that must agree with the proposed oldest age of 200Ma, give-or-take, originally set out and verified by the head of the Ocean Drilling Program. The rock ages are at least an order of magnitude older that 200Ma as has been shown, so that figure is removed from the discussion for the most part.

Investigators have been collecting rock samples from all over the Atlantic basin. Orphan Knoll, 50°23'N, 46°24'W, is a fairly representative sample. Magnetic anomaly 24 lies atop Orphan Knoll, giving an age of 53Ma. DSDP Site 111 established the age at Precambrian metamorphic and granitic rocks, implying a continental origin. Following that Paleozoic limestones were captured that were interpreted to be the nearby bedrock. The feature was dredged again in 1978 with the same results. The primary reason for citing this sample was that the authors of the paper stated unequivocally that the rocks could not have been ice-rafted to their current positions, and this paper was rejected by the reviewers. Possibly they may have thought that, given enough samples, Orphan Knoll would disappear! The paper was finally published in 1995.

The remainder of the North Atlantic basin falls essentially under the same regime. Samples collected by the DSDP corroborate these dates. Leg 37, drillsite 334 core 22 produced an $^{40}Ar/^{39}Ar$ age of 635 +- 102 Ma on anomaly 5 (theoretically 8-10 Ma). Onboard scientists pointed out that "... a remarkable phenomenon..." of the Mid-Atlantic Ridge in the vicinity of 45°N latitude is that the dredge hauls are consistently 74% silicic rocks in the haul, and some areas consist wholly of silicic rocks. One such area, Bald Mountain (45°13'N latitude, 28°53'W longitude), contains radiometric dates from the granite that range from 1,690 to 1,550 Ma (Proterozoic). Gabbroic rocks drilled at Site 334 on Leg 37 (37°02'N latitude, 34°25'W longitude), yielded a radiometric date of 635 +- 102 Ma. Similar rocks drilled at Sites 386 and 387, Leg 43 (on the Bermuda Rise) yielded dates of 671, 381, and 323 Ma (Proterozoic-Carboniferous). [The reader is reminded that dredge hauls do not penetrate the surface, so the ocean floor in the region of these hauls is at least that age.]

On the opposite side of the Mid-Atlantic Ridge in the King's Trough area, gently dipping marine strata have yielded Early Ordovician trilobites and graptolites. Paleozoic trilobites and graptolites, in fact, have been sampled at several localities (Barrie Rickards, pers. comm., 1999). Interestingly, graptolites roamed the oceans from 540 to 320 Ma (Cambrian-Carboniferous). Protolith (original rocks metamorphic rocks form from) ages of several Mid-Atlantic Ridge basalts are 1230 +or- 350 Ma (Proterozoic). At the junction of the rift valley of the Mid-Atlantic Ridge and the Atlantis Fracture Zone (roughly 30°N latitude, 42°W longitude; Figure 12-4), basalts of the Middle Jurassic age (169 Ma) are found. Many investigators now hypothesize that a land bridge existed during a low stand from Europe to North America. At 700Ma the Iceland Plateau seems to have been continental. It was a part of the North Atlantic dam that included Greenland, Iceland, the Faeroes, and Rockall Bank.

The Central Atlantic basin presents another anomaly. Enrico Bonatti of the Lamont-Doherty Earth Observatory (LDGO) has been studying this region for many years. He has concluded that the equatorial region of the MAR has long been recognized as a region of considerable structural complexity that contains old rocks. It is also a region of comparatively low heat flow (<55mW/m^2) along the midocean ridge system where the values are normally >60mW/m^2. Samples from 1992 and 1993 recovered continental flood basalts, sericitic mica phyllite, quartzite, carbonaceous shale, brown coal, and many other rocks derived from continental sources, as well as midocean ridge-type and eugeosynclinal alpinotype rocks. Late Cretaceous-Eocene ages were not uncommon. A 2-300-km wide plateau lies to the west of the MAR, while east of the rift valley it merges into the Sierra Leone basin. This plateau is capped with sedimentary rocks 400 to 1200-m thick, sections which do not thin away from the MAR. The investigators concluded that seafloor spreading was not possible in this 550,000 km^2 region. The omnipresence of so many shallow-water and continental rocks has led investigators to propose that a major E-W structural barrier crosses the Atlantic basin between the Ceara and Sierra Leone Rises at approximately 8°N latitude. Additionally, 140Ma rocks have been recovered at the intersection of the Romanche Fracture Zone and the MAR. The 4-km thick limestones cover a linear distance of 200 km. And, this region, according to the magnetic anomalies, is 0 age. A Soviet study, once again by a group headed by Udintsev (1992), found granitic and silicic metamorphic rocks on the western side of the MAR from the equator to 30°S latitude. His report is summarized in the Fracture Zone section.

St. Paul's Rocks, on the equator, gives up rocks whose approximate age designation is 4.5 Ga. These rocks are prevalent between 30°N and 50°N latitude on the MAR. The same dredge samples have brought up granites and Paleozoic greenstone rocks.

The point is not to be belabored. Summarizing the above, Precambrian protolith ages of many rocks from the North Atlantic basin, especially on-ridge, are abundant. These are mainly in the 1.1-1.9Ga range. What this means is that the actual age of the ocean floor is an order of magnitude greater than that initially proposed.

A plate-tectonic phenomenon that has carried through the years since it's inception is called the "Great Cretaceous Outpouring." This idea was first introduced in 1976 with the discovery of Cretaceous reefal faunas in volcaniclastics of the western Pacific Line Islands and the inferred epeirogenic uplift that led to the concept of mid-plate Late Cretaceous volcanism for the entire area. Similar features were reported from several other areas of the mid-Pacific, and they postulated a "volcano-thermal event which began in earnest in Late Aptian approximately 110 m.y. ago and ended by mid-Maastrichtian time approximately 70 m.y. ago." Stratigraphic analyses of DSDP cores from the central and western Pacific basin confirm that volcanic activity in that region was common, and that two separate major episodes can be distinguished. The Aptian--Cenomanian episode formed of most of the oceanic plateaus in the western Pacific. The younger Santonian--Maastrichtian episode produced several island and seamount chains in the same region. These volcanic events covered an area of about 30 million km^2. This is the gist of the "Great Cretaceous Outpouring." However, oxygen and carbon studies recorded anoxic events in the Atlantic that were coeval with two peak episodes of volcanism (115-110 and 78-72 Ma) in the Pacific along with the deposition of carbonaceous clays in all ocean basins around this time

Apparently the seamounts themselves are impervious to the drill, because no real rocks other than new basaltic glass have been brought back to date. If we use the figure from the preceding paragraph, the seamount takes about 50-Ma to get to where it is now, and that is 110--120-Ma onto that 50-Ma. This is about the 160-Ma range, which is the upper limit for ocean floor age. This means that the ocean floor and the northwest Pacific seamounts had to have formed co-evally within the constraints of the current working hypothesis. This is, of course, not the case, so the answer must be sought elsewhere.

Late Cretaceous volcanism recorded on seamounts and islands from DSDP Site 165 to 100 km northwest of the Caroline Islands, a distance of some 2500 km, belies the hotspot theory which employs a single hotspot of the Hawaiian-Emperor type produced the Line Island chain. "Biochronologic data from the Line Islands indicate that the chain is not the temporal equivalent of the Emperor chain. Volcanic edifices of Cretaceous age are now known to extend from the Line Islands through the Mid-Pacific Mountains to the Marshall Islands and the western margin of the Pacific plate from Japan to the Marianas."

Rock ages from some of the seamounts belie that number, too. The 833-Ma rocks on Tahiti are a case in point. The quartz atop Smoot (Yabe) Guyot is another. The garnets and Precambrian rocks from Detroit Guyot are yet more. And so on... The ages are listed throughout this treatise. Suffice it to say, the age of the seamounts, and the rock constituency, belie the age of the ocean floor at 160-Ma--200-Ma.

The direction of movement of the plates has been theorized to be imprinted by the fracture zones, a concept which allows one to infer an ocean basin's history. The plate-tectonics hypothesis uses the term, fracture zone, to be the inactive section of the transform fault, which forms at a spreading center. Fracture zone trends in the Atlantic Ocean basin were thought to lie mostly WNW--ESE in parallel-to-subparallel straight lines. Those in the Pacific Ocean basin were thought to lie on a WSW--ENE azimuth for the North Pacific and a WNW--ESE azimuth for the South Pacific, with all of them stopping about mid-plate. Where bathymetry was not available, fracture zones could be detected by the presence of disruptions or offsets in magnetic anomalies. Fracture zones do not necessarily form as extensions of transform faults. Additionally, the bathymetry of more than several NNW--SSE-trending fractures and seamount chains is now in the literature, bathymetry for which no satisfactory explanation has been forwarded outside the realm of **ad hoc** "micro-plates." Kashima Fracture Zone (Figure 36) and the San Andreas Fault are examples. Added to the NNW--SSE-trending fractures, three of the WSW--ENE-trending fracture zones have now been proven to transect the Pacific basin, The Clipperton, the Chinook, and the Mendocino megatrends. This provides a checkerboard effect of orthogonally intersecting megatrends.

More specifically, the bend in the Hawaiian-Emperor chain has been repeatedly used to "prove" seafloor spreading changes in plate motion direction at 43-Ma. No plate reorganization occurred at 43- Ma. The assumption is made that another explanation exists for the formation of these seamount chains. No bends exist at 43-Ma. The ages are not in sequence. The guyot tops are not gradually descending according to the bathymetry. In fact, the Tuamotu Ridge and Line Islands appear to be analogous to the Louisville Ridge and Marshall-Gilberts; that is, they were created in response to a leaky fracture zone. That is the reason that the seamounts and islands in the Marshall-Gilbert-Louisville chain, the Tuamotu chain, and the Line Islands were aged non-sequentially. The reason for the age differences in the neighboring seamounts and islands is because a fracture zone is actually a fracture swarm for at least part of its existence, and it is a leaky fracture swarm at that. With a continuously flowing magma channel beneath the fracture zones, a seamount or island may erupt at any time.

Last, the presence of the ubiquitous paleolands on all the ocean floors belies the recent 200-Ma age of the ocean floor. Too many different investigators world-wide have found too much Precambrian crust underlying the fresher basalt to ignore that any longer. The newer maps of the ocean floor ages from Norway, Russia, Australia, and Japan all point to the need for an updated tectonic model.

4. Derivation of Tectonic Model

A. Earth expansion

Realizing that something was amiss with the plate tectonic hypothesis, in 1976 Sam Carey reopened the case for an expanding Earth. The models proliferated, several using nothing more than Earth expansion based on seafloor spreading/ocean floor magnetics with no subduction. This limited the models to conditions from the present back through the Jurassic because, as noted, no magnetic anomalies exist that are older than 180 Ma. Carey, Klaus Vogel, Oakley Shields, and James Maxlow all have a fast-expanding Earth that was only 60% of its present size in the Jurassic. Carey believed the equatorial region to be in a state of sinestral torsion, which was thought to be the combined effect of gravity and rotational inertia. Carol Strutinski changed that to mean an easterly flowing asthenosphere, and that this could also be inferred from Jupiter and Saturn.

Vogel's global models provided what some considered the most convincing model of Earth expansion. He had most of Earth covered by continental crust until the Mesozoic, with the recent crustal movements all being related to radial outward pressing of the continents and the infilling by new oceanic crust from ocean floor spreading at the midocean ridges. His hypothesis does not contain continental drift or subduction, both steps in the right direction. However, the growing radius is in agreement with seafloor spreading. Shields used paleobiogeography: "In Permo-Triassic times, e.g., there are terrestrial biotic links spanning the eastern Tethys and central Panthalassic oceans that don't occur across Pangaea, so explaining these seems to require an expanding Earth."

In order to prove Earth expansion, the moment of inertia constraints must be overcome. An expanding Earth would necessarily rotate slower than a smaller diameter planet. This is called "conservation of angular momentum." In order for this to happen, the lunar tides would have to slow down, which would affect the length of the lunar month. Here the expansionists rallied for many hypotheses. A study of the growth patterns of mollusk shells since the Ordovician shows an Earth year of 447 days at 1.9 Ga decreasing to an Earth year of 383 days at 290 Ma to 365 days at this time. Another parameter, the Devonian coral rings, shows that the day is increasing by 24 seconds every million years, which would allow for an expansion rate of about 0.5% for the past 4.5 Ga. According to the coral growth patterns, this is not enough to warrant a discussion, especially in terms of ocean basin opening, moving continents around, or the dispersal of Pangaea.

Actually, during the past 900 Ma Earth has experienced a slowdown in spin rate according to the NASA space physicists. Even now, this can be detected, and the rotation rate changes in milliseconds per day. This is dependent upon "how the mass distribution of Earth and its atmosphere change from earthquake and the movement of water and air." A further explanation of the spin slowdown reveals that a day was 18 hours long 900 million years ago. Because the Moon and Sun are constantly applying a tidal force to the Earth, the water moves as tides. As that water flows against the ocean floors, it applies a "braking force," essentially a slow-down mechanism, on the planet. In billions of years, this will force the lunar months to increase from the present 27.3 days to 47 days.

The slowdown is explained by the long-term degassing, producing upward migration of the heavier materials into the mantle region from the core, has caused a planetary lowering of the spin-rate, thereby increasing the length of the day.

Chandler wobble and axis tilt also change the rate-of-spin. If the tilt were as much as 54° instead of the current 23.5°, the polar regions would have had the regional warm climates that have produced the fossils being found there with no continental movement. Also, the motion of large air masses can change due to tilt by measurable amounts daily. The change in tilt affects the rate of rotation. Chandler wobble "resides in the natural resonances in the body of the spinning earth due to detailed distribution of mass in its surface, interior, oceans, and atmosphere."

B. Wrench Tectonics

Another form of tectonics has been developed to explain the orthogonally shaped features on Earth's surface. Independently, Karsten Storetvedt of Norway and Howard De Kalb of Hawaii have developed an ideas first espoused by Joe Gilg of NAVOCEANO at Suitland, Maryland (1969) about rhomboid features. This idea was immediately discussed in relation the Sea of Okhotsk, and it expanded from that point.

The two independent investigators discuss the ubiquity of rhomboids, both continental and oceanic. They rely on irregularly spaced degassing of the oceanic crust to produce lateral physio-chemical variations in the mantle. Coupled with intermittent episodes of polar wander and changes in the planetary spin rate delivers the present-day configuration so well-known in the ocean basins. Essentially, these on-again, off-again tectonic forces have throughout time, created step-like latitude-dependent wrenching foldbelts. Thus, the equatorial bulge at the time is totally dependent upon the pole-of-rotation at that time, similar to the force necessary to produce the hotline configuration in the GEOSAT diagrams. Naturally, this change of rotation, the liquidity of the upper mantle, and Earth's spin rate would determine the orientation of the rhomboids. These were of greater significance during the Precambrian that during the Phanerozoic. Many lines point to a faster spinning Archean Earth, which has led to the present-day continental configuration. So, paleo-equators have given much slip-and-slide to the planet in the upper lithosphere, which seems to have been acted upon by almost any hypothesis one wishes to invoke. Anyhow, this one also fits the bathymetry and satellite altimetry, so it is also incorporated into the incipient hypothesis.

Interestingly, Storetvedt's Figure 9 delineates a Eocene equator which shows the exact azimuth of the megatrends going NE through the Atlantic basin (Figures 10 and 11). As a matter of fact, many of the proposed lineaments agree with the later data in both of the hypotheses. At the Proterozoic boundary a major change occurred in global tectonics. A new relative equator passed from the Arctic Canadian and Labrador seas along present Central and South Atlantic. Orthogonal to that equator were created the Grenville Province, Ellesmerian-North Greenland zone, and the pan-African and Brazilian mobile belts. During the Middle Ordovician another polar wander occurred as the paleoequator ran across the northern Atlantic. This formed the basis for a global wrenching that produced the Appalachian-Caledonian foldbelt.

This feature would seem to be the same as the Cladonian Magatrend discussed in the chapter on megatrends. Subsequently, the equator has been in different positions across Europe, leaving tectonomagmatic belts in its wake, getting younger to the south.

C. Surge tectonics

In order to continue this discussion, a more robust tectonic hypothesis must be interjected so that the above examples may be placed in a unified framework. That hypothesis is called "surge tectonics," and it is essentially based on the hot lines/megatrends.

Using the data available during the 1980s, a group tried to incorporate all of the real data into a more robust explanation as to Earth's tectonic history. They assumed that rocks exposed at the surface should cool, and that was the start. The net ever-diminishing Earth's radius and increase in rotational acceleration, along with the four cycles of tectonics are combined in the surge-tectonic hypothesis. Evidence exists in the lithosphere of pipe-like, tubular, or lens-like hot bodies that are interconnected. As we have seen, this has been known for the past 25 years. These hot bodies underlie foldbelts, rift zones, and strike-slip zones. The hot, lenticular bodies are called surge channels, occurring at depths of between 40 km and the top of the asthenosphere. The hot bodies move as Earth contracts in a sequence called the geotectonic cycle.

The geotectonic cycle is described as undergoing periods of alternating taphrogenesis and tectogenesis whereby: (1) In the tectogenesis process, contraction of the lithosphere is always occurring because Earth is always cooling. (2) The overlying lithosphere is already cool, so it does not contract. Instead, Earth's basal circumference decreases in discrete pulses by large-scale thrusting along Benioff zones in the lithosphere. This process has been called subduction in the plate-tectonics hypothesis, and that term is sufficient still. (3) The discrete pulses of large-scale thrusting along the Benioff zones occur when the underlying dynamic support of the lithosphere fails. The weight of the lithosphere overcomes the combined weight of Benioff zone friction and the asthenosphere. Hence, tectogenesis is episodic. (4) During periods of lithospheric stability, or anorogenic intervals, the asthenosphere volume increases slowly as the strictosphere, or mantle, radius decreases and decompression of the asthenosphere begins. (5) Decompression is accompanied by rising temperature, increased magma generation, and lowered viscosity in the asthenosphere. The asthenosphere gradually weakens during the time intervals between collapses. (6) During lithosphere collapse into the asthenosphere, the landward sides of the Benioff zone obduct the ocean floor. The lithosphere buckles, fractures, and founders. Enormous compressive stresses are produced by this action in the lithosphere. The position of the surge channels is partially controlled by the presence of Benioff zones. (7) When the lithosphere collapses into the asthenosphere, the asthenosphere-derived magma begins to surge in the overlying surge channels intensively. Where that volume exceeds capacity, and when the compression in the lithosphere exceeds the strength of that lithosphere directly overlying the surge channel, the surge channel roofs rupture along-strike in cracks that comprise the fault-fracture-fissure system generated before the rupture. Rupture is bivergent and forms continental rifts, foldbelts, strike-slip zones, and midocean rifts. These ruptures, called kobergens, are mountain ranges and midocean ridges, and the identifiers can be used interchangeably from a geomorphology standpoint through terrestrial/marine settings. (8) Once tectogenesis is complete, another geotectonic cycle sets in, usually within the same belt.

Because most of the earthquakes lie above 100 km, the first assumption is that long, linear features with elevated heat flow and earthquake activity lie at least that deep in the lithosphere. This places the feature just above the asthenosphere. Just as lava flows, so does the magma occupying these features, or channels as they are called.

Secondly, when a body cools, it contracts in most physical cases. The Earth is such a body, and it has been cooling for the past 4.5 Ga. That is why Earth has a hard, outer shell. This is called the lithosphere. Because Earth is not dead, this hard outer shell covers the heated inner sphere. Magma rises from this inner sphere to fill the asthenosphere with molten rock. The asthenosphere is not everywhere the same thickness, so that channels may form both above and below the asthenosphere proper. The channels which form above are called surge channels in this hypothesis, and that introduces what we believe is taking place all over the world-channels form a web in the lithosphere.

Third, the channels intersect. Where they intersect is similar to a traffic roundabout. Only so much space exists for the magma to occupy. When two or more channels intersect, this provides an excess amount of magma to that particular area. So, the magma, already heated, superheats the area around the intersection so that it becomes semi-molten. The magma swirls around depending upon the laws of whichever hemisphere the observer is in, and creates a roundabout effect. The features are called vortex structures. Some are upwelling vortices, and some are downwelling. The upwelling vortices are the rises and plateaus on the ocean floor.

Three forces drive surge tectonics. That the Earth is cooling is a given, and that is the first driving force. The second is Earth's rotation coupled with drag along the outsides of the surge channels. The process involves differential lag between the already cooled lithosphere and the strictosphere. The third driving force is gravity. The outer shell, or lithosphere, is already cool. As the strictosphere cools, it recedes from the underlying support provided to the cooled lithosphere. The surge channel empties, and gravity makes the roof (lithosphere) collapse over the surge channel. The collapse is analogous to stepping on a tube of toothpaste. Any magma in the chamber is forced to surge laterally, and a surficial expression in the form of a fracture zone is formed.

The process has been going on ever since the world began, or at least since the Proterozoic according to the many rock samples shown.

The lateral movement of the magma in the surge channels is addressed. Mantle diapirism had received very little attention until the late 1980s. The plate-tectonic concept cannot account for juxtaposed compression and tension regimes. Surge tectonics, on the other hand, requires such a stress field. This phenomenon is nowhere more evident that in a cooling Earth. As Earth cooled, it did and does so downward from the lithosphere. This means that the lithosphere is in a state of compression constantly and ubiquitously. The region from the strictosphere to the center of the Earth is in a state of tension. The asthenosphere between them is the level of no strain, so continued cooling drives the asthenosphere deeper over time.

The asthenosphere alternately expands during times of tectonic quiescence and contracts during tectogenesis; it pulsates. These pulses are what defines Earth as a giant hydraulic press. Pascal's Law states that pressure applied to a confined liquid is exerted undiminished throughout the fluid in all directions and equally upon all areas at right angles to it's interior surface. In layman's terms, this is the guts of the surge-tectonic hypothesis. The gravity-forced collapse of the overlying lithosphere, the roof, into the asthenosphere, or surge channel, drives the hydraulic press. The interconnected asthenosphere and surge channels are the closed vessel, and the magma is the liquid in the vessel.

Where the volume of magma in the channel exceeds the capacity of that channel, and when compression in the lithosphere exceeds lithospheric strength in the area above the channel, the channel roof ruptures along the pre-existing fault-fracture-fissure system generated by Poiseuille flow before the onset of the event. The bivergent rupture forms continental rifts, foldbelts, strike-slip zones, and midocean rifts; kobergen structures. The importance here is that the end product, or tectonic belt, depends entirely on the strength and thickness of the lithosphere overlying the surge channel. Because tectogenesis effects the entire tectonic belt according to Pascal's Law, the event may be world-wide.

Reducing all of the above to plain English, the rise-type, positive gravity anomaly features such as seamounts, plateaus and rises, aseismic ridges, volcano arcs, and midocean ridges are all products of taphrogenesis where the surge channels are filling and flowing. The region over the surge channels is expanding somewhat due to the excess heat being brought to or near the surface, but this is no way to be construed as seafloor spreading or Earth expansion. During tectogenesis, the surge channels are emptied, Earth's diameter decreases, and compression features such as trenches, ocean floor fabric, and fracture zones are formed. The compression is a unidirectional move in a contracting Earth, as the entire body will never be that size again.

Earth's rotation is always easterly, so that is necessarily the alignment of the tectonic belts, or surface features. The lineaments change directions at discrete intervals. This is because Earth wobbles on its axis. The primary lineaments can then give a fair representation of the location of Earth's poles at the time the feature was formed. This will be nowhere more evident than the intersecting fracture zones in the Pacific Ocean basin.

The tools for mapping the asthenosphere flow structures, or vortex streets, are (1) GEOSAT data, (2) concentrations of epicenters, (3) belts of microseismicity, (4) seismic reflection studies show cross-sectional depth, size, and location of flow structures, (5) seismic refraction studies map the velocity and activity of flow structures, (6) linear-trending surface basalts yield the ages of flow structures, and (7) perturbations of the magnetic and gravitational fields may indicate movement of subterranean fronts or movements in large scale flows.

These phenomena were collectively mapped in the 1980s (Figure 58) by John Woodhouse and Adam Dziewonski of Harvard University to show the hot and cold regions in the surge channel depth range, from 50--150-km, which have been determined from seismic refraction velocities. In the interest of clarifying the surge tectonic hypothesis, this figure is explained. First, the proposed driving force of plate-tectonics, convection cells, cannot exist as illustrated by the cold regions under the oceanic plates, especially in the central Pacific. The presence of fracture zones precludes the contiguous plate concept. Without a contiguous plate, no contiguous convection cell exists. A heated convection cell cannot exist under the colder central Pacific basin. The fact that the central Pacific basin is cold is directly related to the presence of the Benioff zones, which act as a general barrier to the eastward flow of the surge channels. This phenomenon is nowhere more evident than in the western Pacific, where the trench system from Kamchatka south through the Mariana trenches and the southern Tonga/Kermadec trenches are major dams with the exception of the breakout channels. The older, deep-cratonal roots on most continents also act as dams. The street being discussed generally follows the hot, slow, equatorial dark zone.

Taking all of this book into account, we can now undertake the route of portions of the present-day surge channels in the context of basins rather than plates. One has merely to find the route between trunk channels to keep the model moving. We will not only do this, we will carry a channel around the world by expanding on a previous exercise which outlined the first world encircling vortex street. Words used interchangeably in this discussion will be vortex street, geostream, and active surge channel, be that a trunk, breakout, or feeder channel.

While the surge hypothesis does not subscribe to polar wander, this must be the case because of the wide variation in megatrend azimuths shown by all of the ocean floor data. However, wrench tectonics otherwise agrees wholeheartedly with surge tectonics. What we have done herein is to combine at least those three different hypotheses into one which is more robust than any of the three standing alone. It is serendipitous that many different hypotheses, coming from many different directions, all arrive at the same conclusions, and the real ocean floor data substantiates all of the included hypotheses.

All of the outlined models include at least three phases of Earth evolution. The continental shield/granite stage was during the Archaean Era. The original continental crust was created by calc-alkalic magmatism before the Mesozoic in a series of steps. These in order were geosynclinal, orogenic platform, continental rift, and block tectono-magmatic activization. The Proterozoic and Paleozoic Eras were called the transitional stage. The Mesozoic and Cenozoic are called the basaltic stage. In some, basalt is replacing granite from the midocean ridges by seafloor spreading. In others basalt is covering the continental crust which remains in place. In yet another, the continental crust is being converted to basalt.

A 2002 conference by the Scientific Council for Earth's Problems was held by the Russian Academy for Sciences (**Geological Structure and Origin of the Pacific Ocean**, edited by B.I. Vasil'yev, 2005). The results reinforced many of the data being shown herein. One item stands out among the rest. Boris Vasil'yev made an updated map of the Atlantic Ocean floor age, this time based on rock ages instead of magnetic stripes. This is the first time an ocean basin has been defined in the same geologic terms that are applied to maps based on continental numbers. The map does not look anything like the maps based on magnetics, totally refuting the ideas for this basin put forth by mobilistic reconstructions. Interestingly, the trenches are found to be very young. This study, regardless of others that have shown the same thing, shows the idea of the magnetic measurement, as it is used now, to be totally fallacious. And, this has been known since the first studies done in the 1970s refuted the 1950s ideas.

At this same conference, V.T. Frolov presented facts that clearly demonstrated that no possibility exists for the prior existence of a Paleozoic Ural Ocean, thereby refuting previous work by Sengor. The region has not undergone any significant contraction. Igneous and sedimentary rock studies, along with the geologic formations and styles of folding and faulting, do not confirm extension for hundreds of kilometers. Neither do they confirm subduction, accretion, or tectonic piling. The study does confirm that the Paleozoic Ural Mountains were dominated by vertical tectonism.

5. Vortex Structures

Added to the above mixture, a newer type of feature is introduced, that of vortex structures. The two types are 1)upwelling and 2) downwelling. The upwelling variety has already been introduced, just by another name. They are the plateaus and rises which occur at the juncture of two or more channels. One indication of the presence of an active surge channel is microseismicity. Another indicator is structural trends.

Establishing the presence of surge channels is important because of the appearance of the vortex structures in the bathymetry. Their existence is predicted wherever surge channels intersect in the surge-tectonic hypothesis, this intersection being much the same as a highway roundabout. At the point of the intersection there is too much magma for the channels to hold, and a vortex forms.

One of the hindrances to the temporary passage of a surge channel is a continental craton. Passive margins on the eastern side of a basin act as Benioff zones do on the west; they dam the efforts of the ever easterly-flowing surge channel. Where this occurs at surge-channel intersections, vortices again appear. An example in the NE Atlantic basin shows the combined Oceanographer/Hayes fracture swarm coming in diagonally from the lower-right corner of the figure. The Azores-Gibraltar Ridge is the primary feature from the bottom. MOR fabric appears to the left of that. The circular feature on the top of the figure is the Horseshoe Seamounts, a perfect vortex. The vortex has been partially infilled by sediments from the Tagus River in Portugal, building the Tagus Plain. The Horseshoe Plain is on the top right (Figure 59).

Vortex structures were previously noted in 1929, but were not addressed by the plate-tectonics hypothesis. Many terrestrial vortices were known, such as flood basalts, the Pannonian Basin, Lake Victoria, the Dasht-e-Lut depression in Iran, the Deccan Traps, and the Columbia River flood basalts. They have been shown to be large, ovate or oval-to-circular regions up to 1000 km in diameter, beginning as eddy-like structures. In general vortex bathymetry shows a whorl pattern or across-basin fractures dominated by a major depression. From a geophysics standpoint, vortices are also recognizable by excessive micro-earthquake, volcanic, and hydrothermal activity, high heat flow, and negative gravity anomalies. As an example, the North Pacific Vortex resides in the region of the Hess Rise/Emperor Fracture Zone/Emperor Seamounts. Bathymetry and gravity data show at least three eddies where the Mendocino/Surveyor fracture swarm offsets the Liliuokalani Ridge and the Emperor Fracture Zone.

By the late 1960s, a major problem occurred because vortices did not fit into the plate-tectonics paradigm and were rarely discussed. The bathymetry in the literature has since shown the North Fiji Basin, the Easter Island "micro-plate" (see cover for earthquake outline of this vortex), and the South Adriatic basin as vortex structures. This has led to their recognition as geomorphologic entities. While not recognized as such, Dan Walker, a retired geophysicist from HIG, has published a book in which he waxes eloquently about the many possibilities for the

Easter Island vortex (**Volcanoes, El Ninos, and the Bellybutton of the Universe**)

Meyerhoff and others have found that vortex structures are ubiquitous. In fact, eddies are prevalent in most vortices. As an example, the North Pacific Vortex resides in the region of the Hess Rise/Emperor Fracture Zone. Bathymetry and gravity data show at least three eddies, one of which is included below. The intersection of the Musicians Seamounts and an unidentified feature cutting cross-trend is an incipient vortex in the making. Now the bathymetry of the Aegean Sea is added to the list. The GEOSAT data base proves the existence and the location of vortex structures.

Vortex structures exist as fracture zone offsets such as small vortices called overlapping spreading centers. Channel offsets are eddy structures. Vortex structures themselves are the largest. Eddies are represented by the three sub-vortices in the North Pacific Vortex, the San Clemente fault zone, the San Francisco Bay, and many others. Examples of both an incipient vortex and a mature vortex are presented as an aid to understanding the evolution of vortex structures. The incipient stage of vortex formation is shown by SeaBeam bathymetry off the coast of California. The incomplete whorls at 45° to both of the regional trends, amply demonstrates how a counterclockwise vortex begins.

Skipping northward to a multiple fracture swarm intersection (Figure 18), the juncture of the (1) Emperor Fracture Zone, (2) Hess Rise, (3) the Liliuokalani Ridge, (4) the Musicians Seamounts, all of which trend northwesterly, and the (5) Mendocino Fracture Zone and the (6) Pioneer Fracture Zone, both of which trend southwesterly fully demonstrate the topic. Intermingled into that morass, we find a completely mature vortex (7). Thus, orthogonal fracture zone intersections are home to either upwelling features, such as plateaus and rises, or downwelling features, such as vortices. This vortex structure is an eddy in the larger North Pacific Vortex.

6. The Mediterranean basin and Southwest Asia

The tectonic history of the Mediterranean basin is in a state of flux; nobody seems to be able to put the region in a plate-tectonics framework, though many have tried. Most recently the central basin has been shown to be in collision to the west of Sicily, and to the east oceanic crust is subducting the Ionian Sea; that is, the Pelagian block and the Calabrian arc. From the backarc Aegean region up to the North Anatolian Fault Zone the compression and extension are taken up by: (1) compression along a mean direction of 140°, rate 1.1 to 2.8 mm/yr and (2) extension along a mean direction of 050°, rate 9 +or- 8 mm/yr. About 44 Ma the African lithosphere drove north into the Iberian lithosphere, which was a separate regime at that time. A piece of the African plate split off to form the Apulian plate to the east, containing Greece, Albania, Yugoslavia, and western Turkey. The Eurasian plate was to the north. By 9 Ma the plates and continents all piled up the present-day configuration for this region. The ever-widening Atlantic Ocean is pushing Iberia and Africa towards Apulia, with Africa subducting Iberia. A collision zone was created along Italy, marked by the Appenine Mountains. This gives us an east-west subduction zone for the western Mediterranean and a vertical collision zone for the central Mediterranean. In the eastern Mediterranean, the Apulian plate moved north and thrust up the Alps. From the south, the African plate is pushing oceanic crust northward at the rate of 5 to 8 mm/yr, and this crust is subducting the Aegean Sea region, leaving a continental fragment in the form of Crete and ophiolites on Cyprus. Subduction here is pulling away from western Turkey, which is being driven west by the Arabian plate collision. The thickness of sediments on the Mediterranean ocean floor precludes the bathymetric observation of trenches, so we will have to switch to geophysics to study this region. Earthquake seismicity shows that no trenches exist in the Mediterranean. Also, no plates exist in nature.

If this is a subduction zone, it is atypical. No organized trench axis exists as such, the bathymetry and gravity showing an outline of unorganized lows to the south and east of Crete instead. A 4380-m low there and another southeast of Rhodes are the predominant low trends. Otherwise, there is nothing on the one-minute grid that even resembles a trench. The gravity data show a low zone in the same spot.

The region has been analyzed in a surge-tectonics framework by Bruce Leybourne using earthquake data from the Italian Telemetered Seismological Network of the Instituto Nazionale di Geofisica. From 1986 to 1990 the surge channel moved northwest. The second earthquake swarm shows a breakout attempt in January 1986. The third dated earthquake swarm in sequence shows that the surge channel moved northwest through the Apinnines in October-December 1986. The fourth dated earthquake swarms show two forks. The one is the northern extension, on course, of the primary surge channel which occurred in July 1987. The other represents a breakout attempt during July 1987, two days earlier. That breakout having failed, the system tried two more times to break out in September 1987. Finally, in April 1988, a large earthquake measuring 5.3 (Richter scale) marked a successful breakout attempt in the central Adriatic region. This carried the breakout channel to the eastern surge channel moving southeast through Yugoslavia, Albania, and Greece. The breakout completely opened with the swarm from October 1989 to October 1990, creating a typical, incipient, downwelling vortex structure for the Adriatic region (figure based on Table VII).

Table VII. EARTHQUAKE DATA FOR THE MEDITERRANEAN SEA REGION.

(data from the 1990 NGDC earthquake CD-ROM)

Earthquake depth (kms)	Quantity
0-4	35,927
50-99	913
100-149	436
150-199	192
200-249	23
250-299	54
300-349	23
350-399	6
400-449	4
450-499	5
500-549	1
550-599	0
600-649	3
650-699	0
unknown	196

Vortex structures originate when a surge channel encounters an obstacle to its passage. An intersection with another surge channel would be a prime example, which is exactly what is happening in the Aegean Sea region. Vortices range in size generally from 300 to 800 km in diameter and are easily recognized by many characteristics: disorganized bathymetry dominated by a major depression, fracture patterns arranged in concentric circles, and intense concentrations of micro-earthquakes with high tensile stresses in the middle.

The evolution of this region is important because of the need for India to be in place for at least the past 600 Ma. Using bathymetry, geology and geophysics, paleobiology, seismotomography, and stratigraphy the determination has been made that epicontinental Africa, Apulia, and Arabia were united almost continuously during the entire Phanerozoic (570 Ma to the present). The seismic reflection and refraction studies show that a 21--26-km thick sedimentary crust overlays a continental crust in that region. This passive regime ended with the Late Cretaceous-Paleocene Alpine compression. The implications here are twofold. First, the proposed spreading, splintering, and closing by plate tectonics of the Mediterranean Basin are wrong. None have existed, so the consensus model from 44 Ma to the present is not based on fact. Secondly, the implications will become immediately apparent in the section on the India collision.

The surge channel flows eastward from the Mediterranean, in part along the Anatolia Fault and in part along the Zagros Crush Zone and the Makran convergent margin (Figure 52). From there it is deflected to the north by either the Ornach-Chaman Fault or the western pinch in the Afghanistan Gap and forms a counter-clockwise vortex at the Dasht-e-Lut, a huge depression surrounded by mountain ranges (Iran). The Dasht-e-Lut is characterized by high heat flow, active volcanism, high seismicity, ophiolite belts, kobergens, and other surge channel features.

In a GSA article on middle eastern tectonics a group headed by D. Seber (1997): "A consequence of the collision between the Arabian and Eurasian plates was the development of the high Turkish and Iranian plateau in...Although the mechanism that holds up this high plateau is not well

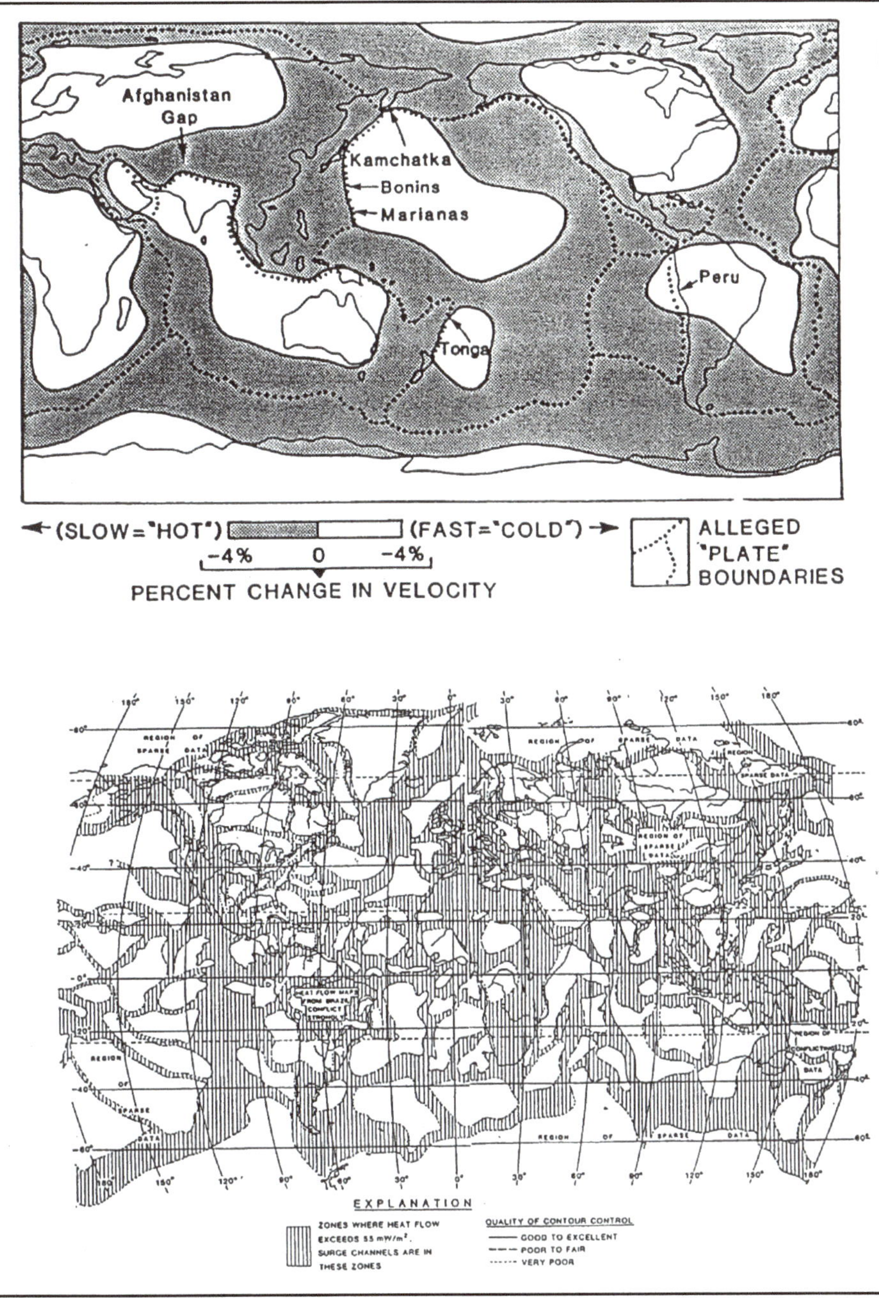

Figure 58. Early 1980s Earth heat flow model by Woodhouse and Dziewonski (top). Refined model based on earthquake and volcano data by Meyerhoff in the late 1980s (bottom).

understood, extensive volcanism and strong seismic shear wave attenuation in the mantle lithosphere beneath the plateau suggest that a thermal component is required..." This is the Dasht-e-Lut and was explained as being supported by surge channel activity.

7. Indian Ocean region

The oldest mafic volcanics in the Himalayas are 2.51 Ga. If it has not already been done so, the splayed surge channel traces, a product of flow, totally negate any arrival models after that time. Also, the biota of southern Laurasia and northern Gondwanaland are 80% the same. The Indus-Yarlung suture zone is commonly listed as the boundary between the southern Gondwana taxa and the northern Tethyan taxa. This zone extends through New Guinea, passing through the middle of the study area. An intercalary zone extends Gondwanan taxa into Siberia, Manchuria, and China. At the same time, the Tethyan taxa extend to New Zealand, Australia, southern India, and Saudi Arabia on the south. This intercalary zone has existed at least back through the Phanerozoic. The zone ranges from a few-hundred to five-thousand kilometers wide. The islands surrounding the Banda Sea have been in place since at least the Jurassic. India is ringed by tropical Tethyan Jurassic and Cretaceous marine sediments that, close to the southern tip of India, contain tropical reefs of Urgonian (132-124.5 Ma) facies. The presence of tropical Tethyan faunal reefs places India in situ since at least the Early Jurassic (208-178 Ma), certainly before the Gondwanaland breakup at 180 Ma, or when India was adrift in the middle of the Tethys Sea during the Eocene.

While still in India, an onshore vortex structure is discussed. The Deccan Traps are a large vortex formed by surge channels flowing WNW--ESE from the north and intersecting the Chagos-Laccadive Ridge flowing south from the north, colliding in the region of Bombay, India. The dike swarms follow the Narmada-Son rift (one of the eastward-splaying horsetail surge channels in paleo-India) to join the Rajmahal traps. Based on the earthquake data the northern surge channels are in a period of quiessence. Only the southern branch, the Siwalik Hills and out onto the ocean floor, is still active.

One item of interest here is the two large-sized ridges in the Indian Ocean basin, the Chagos-Laccadive and the Ninety-east (Figure 13) ridges. The Chagos-Laccadive Ridge appears to be a breakout channel at the Kashmir cusp, and the Ninety-east as a part of the Tibetan cusp. They appear to be the crutches holding India in place. Were the ages to get younger in a southerly progression, they would both be analogous to the Emperor Seamount chain and the Palau-Kyushu Ridge. This would mean that the older start of the channel was to the north. The features passing south from the Kashmir Cusp get relatively younger from Pakistan to the Deccan Traps to the Laccadive and Reunion tracks, so that part of the equation is right. However, the Chagos-Laccadive Ridge, originally thought to be a north-flowing feeder surge channel, is just that. It formed the Aravalli, Vindhya, Satpura, and Ajanta Ranges. The Chagos-Laccadive channel also fed the Eastern Ghats, eventually moving offshore in the mid-Miocene. So, the two surge channels which intersect to form the Deccan Traps are unrelated, flow in two entirely different directions, and have the appearance of being one. As for the Ninety-east Ridge, not enough is known about it to make such a definitive statement.

The Chagos-Laccadive channel has created havoc with the bathymetry of the Central Indian Ocean Ridge. Fractures there seem to have been "dragged" into their present configuration . For this to have happened, the eastward flow of magma had to have been obstructed. That obstruction is the Kerguelan Plateau, which has been found to be a continental land mass. The reason for the drag effect is explained. The eastward flow of magma in that region is the surge channel under the Archaean high in south Africa. It enters the Indian Ocean basin at Madagascar and continues to Reunion Island. There it forms a vortex at the Southwest Indian Ridge/Central Indian Ocean Ridge, and continues northward in the route already outlined. The channel passing under Madagascar intersects the Chagos-Laccadive Ridge surge channel, so we have the beginnings of another roundabout, or vortex structure. That always creates a drag effect on the bathymetry as has been repeatedly shown herein.

The Madagascar/Reunion channel also runs to the east as part of the Indoysian Foldbelt, which is mapped by heat flow. The older, inactive portion is the Diamantina Fracture Zone, which skirts southern Australia.

In a total turnabout, a recent drilling cruise by the ODP, sponsored by the NSF and conducted by the University of Washington, reveals that many zircons have been recovered from samples of the Indian Ocean floor. The age is reported to be 2.5Ga, and it is admitted that ocean floors are an order of magnitude older than we have been led to believe for this past 60 years!

8. Southeast Asian region (with input from Bruce Leybourne and Art Meyerhoff)

Many explanations exist for the evolution of the SE Asia landmass. The option here is for a far simpler explanation, one provided by Art Meyerhoff. The northern branch of the Central Pacific surge channels arises east of the Afghanistan Gap in a pinch-and-swell situation, where the Himalayan Mountains rose above the now-extinct surge channels in an orogenic phase, a period of tectogenesis. The principal continental land masses had acquired their present size and shape since the pre-Sinian, a fact which is verified by the existence of the horsetail-shaped, eastward fanning faults of central Asia.

Having India **in situ** for all of its history helps explain much of the tectonics of this region. Because the splayed surge channels were in existence for the 400 Ma preceding the end of the Ordovician, the western Pacific Benioff zones must have necessarily already been in place. During the Silurian-Devonian the site of the Indonesian archipelago seems to have consisted primarily of deep-water troughs that surrounded shallow-water, small continental platforms. Widespread germanotypic tectogenesis and epirogenic movements took place. The trench system stretching from the Andaman-Nicobar Islands to New Guinea was also in place. The surge channel came SE out of India in a series of five splayed channels, proving that India was in place by the Permian at the very latest.

A time-sequenced series based on actual rock ages shows how the present-day Pacific surge channel evolved in Southeast Asia. The island arcs are the only remaining questions. Most remain unmapped in any detail in this area of some 10,500,000 km^2. Sampling has given at least a Paleozoic basement to the Philippine Islands, Proterozoic granitic clasts in western New Guinea (1,250 Ma) and northwestern Irian Jaya, 1,700--1,500 Ma on western Malay Peninsula. Without sufficient sampling, the statement is made that much of southeastern Asia is underlain by Proterozoic basement, making it coeval with the more well-sampled mainland.

From all structural appearances, the Banda Sea region is an active portion of the currently inactive Indonesian vortex; a possible eddy. Sulawesi appears to be the dividing line between the two. Thus, the Banda Sea vortex is defined by the surge channel traveling eastward under the island arc and north of the trench/troughs. It is deflected north, some of which continues to the east through the Aru Trough to Irian Jaya. Another arm goes northerly to the east of the Weber Deep, only to be deflected to the west by the arriving surge channel under the Palau-Kyushu Ridge, which intersects here at Halmahera. That portion is met by the surge channel passing along under the North and South Sula Faults. That fault system is an eastward extension of the southerly traveling Palu-Koro Fault System, and this portion completes the circle for a vortex structure. Sulawesi is a K-structure, and the faults are remnant surge channels that have collapsed on themselves.

A more thorough analysis of the earthquake data as to the times of the events gives not only the direction of flow but also the amount of time required to make the complete circuit around the Banda Sea. The Banda Sea vortex is partly fed by a counter-clockwise mantle stream which is pulling fragments of continental Australia into the vortex as the mantle upwells from a depth of 21 km under the outer arc near Kai to 14 km under the Weber Deep and to a depth of 8 km under the North Banda Sea. The mantle rise of 7 km between the outer arc and the Kai Islands exactly equals the depth of the Weber Deep, 7 km from the outer arc high. This is isostatic compensation at work. The Weber Deep extension is caused by the momentum of the mantle stream as it turns northward, which slows the stream velocity and compresses the inner side. The compression causes higher temperatures on the inner side, and this in turn feeds the volcanic pipes in the outer arc. Loss of momentum and excess heating cause the magma to rise.

Discussions of the 1972 Banda Sea earthquake suggest seismic sub-events propagated unilaterally along dip and bilaterally along strike forming a fan-shaped rupture area. A local northward convex plane inferred from the relative strikes of the major seismic sub-events contradicts the overall configuration of contorted, subducting slab of the Australian Plate. The

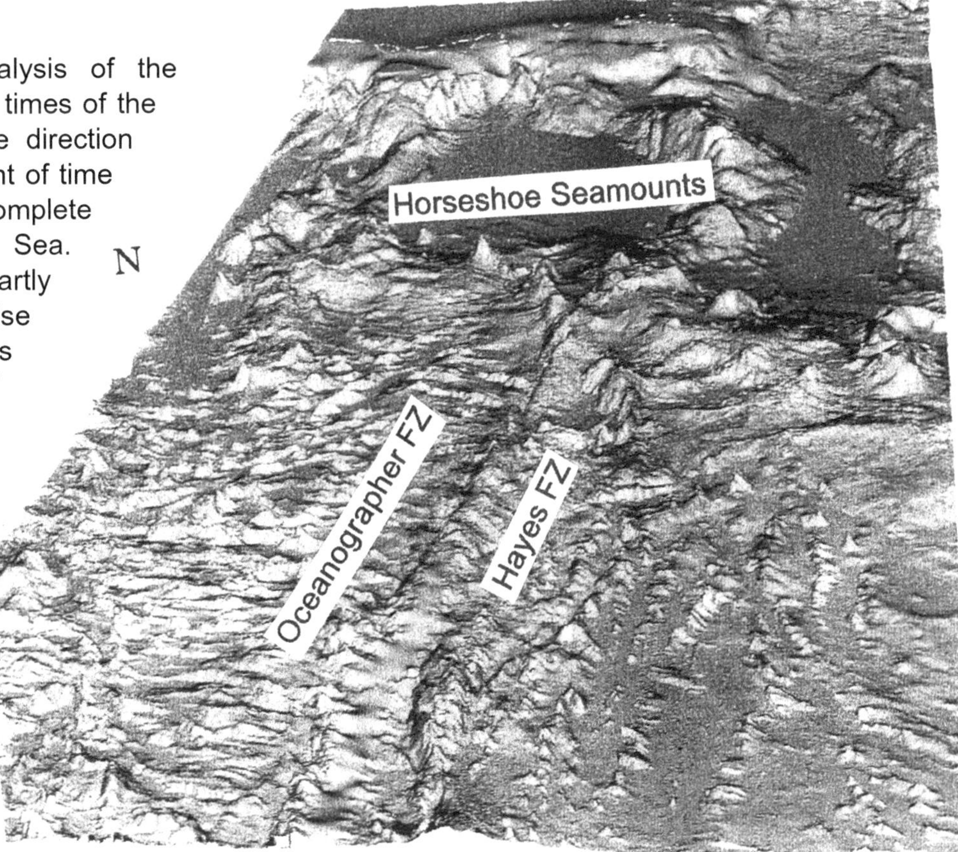

Figure 59. The eastern portion of the Hayes/Oceanographer Megatrend showing the Horseshoe Seamount vortex structure off the coast of Portugal. Once the magma leaves this vortex, it apparently passes through the Iberian Peninsula and into the Mediterranean in the form of the Balearic Isles. There it goes to the vortex off western Italy.

vortex-surge model for this particular case requires a vortex shear which has a schematic similar to the plate model, but the three-dimensional dynamics would compliment earthquake studies by vortex rotation and upwelling along strike with dip in a northward convex, fan-shaped surface (i.e. vortex wall mass) and a decoupled process along the geostream strike after the vortex passes. This model indicates Timor would be rising in rhythm to vortex approach and expansion phase even before it was ripped from the Australian continent. The vector strike-slip component of this process may be up to 75 cm/yr as determined by magnetics or higher up to 195 cm/yr as determined by paleo-bathymetry studies as previously discussed in these cases. Within the new tectonic vortex model accretionary complexes of continental and oceanic crustal fragments sheared by the vortex float atop the geostream to the south. To the north, they are further fragmented within the backarc basins and subducted, or accreted to new micro-continental masses left in the wake of the vortex.

Vortex inflows may be broadly classified as crustal, shallow-, mid- and deep-level mantle inflows and may further be divided regionally. In the northwestern quadrant, surface expression portrays Eurasian crust and shallow-mantle convergence toward the vortex with Pacific mid-level mantle, obscured somewhat by shallow-level outflow to the north. This is a region of turbulence and back arc basin crustal subduction and fragmentation. In the northeastern quadrant deep incoming Pacific mantle converges toward the vortex, except the mid-level mantle outflow through Papua New Guinea. As this outflow shallows eastward, it generates an opposing high-velocity counter flow back toward the Banda Sea Vortex exemplified by arc building along a vortex street en route to New Zealand. Deep mantle upwelling from under Australia wraps continental crust into the vortex from the southeastern quadrant and merges with the rotational component of the Eurasian Upper Mantle Stream in the southwestern quadrant. Thus cyclonic mantle inflow from all directions, generate higher velocity upper mantle outflows permeating the pacific basin generally around the perimeter, but also converging along the East Pacific Rise.

Incoming mantle streams flow in at depths greater than 30 km and converge from three directions, corresponding to the triple junction plate boundaries, and deepen clockwise. They enter the vortex by up-ramping cyclonically within the extension of the Weber Deep. The topography of this upwelling mantle interface would roughly reverse image the bathymetry of the isostatically compensating Weber Deep. There is approximately 7 km of mantle upwelling from 21 to 14 km, for 7 km of bathymetric fall. Mantle inflows are analogous to feeder bands entering the eye-wall of a tectonic tornado or hurricane. Counter flows for each feeder band are generated within the vortex from anatexis of the mantle. These upwelling magmas are extruded at higher depths above and opposing incoming flows. The shallowest incoming mantle stream (Eurasian Upper Mantle Stream) comes in directly from the Himalayas under the Indonesian Island Arc. Counter flowing magmas are extruded along the volcanic arc at the surface as High Velocity Outflow, the bulk of which flows westward within lighter density geostreams. There is some confusion as to classification of these geostreams as asthenosphere counter flow, which is less dense, higher elevation, higher velocity and opposite to upper mantle streams, which in some schemes may also be termed as asthenosphere. Within the surge tectonic hypothesis the upper levels of these complexes are termed surge channels. This is a simple construct, in actuality several layers of opposing flow regimes radiating along triple junction boundaries are evident. The opposing streams are density/depth dependent reaching levels of neutral buoyancy, much like opposing jets found within the inter-tropical convergence zones of the world's oceans and mid-levels of the atmosphere. Seismically these flow regimes are separated by intense earthquake activity along the Benioff Zone were magma exchange occurs.

In an exact duplication of tectonic events, the emission of the upper-level surge channel rotating clockwise out of the vortex at Seram Island creates a similar scenario. Beginning just south of Seram and continuing back towards Indonesia, a K-structure is left in the wake. This structure may be an expression of surface boil. The south end of the K-structure is centered on the upramping side of the incoming mantle stream, which is transporting continental fragments southeastward. The north side has a small, vertically oriented component which opposes a counter-current centered on the K-structure. This action is causing the K to grow to the north. To the west of that is a counter-clockwise rotating basin adding continental fragments to the northern arm of the K-structure and pulling fragments into the basin for surge-tectonic-style backarc basin subduction. This micro-continent is trapped in the vortex flow and moves with it until such time as flow ceases. In that contraction phase, the micro-continent is deposited as the vortex flow moves eastward, leaving this fragment in it's wake.

Two K-structures exist within the unique tectonic setting of Banda Sea region. The southeastern leg of Sulawesi (the larger K-structure) is fragmented and tugged southeast by the upwelling mantle incoming northward from southeast Asia. The southern K-extension straddles this mantle stream. A series of small opposing eddies act as a trapping mechanism of the K-structure. Another way to perceive this phenomena is as the expression of the surface boil of the upwelling magma. Under the K-structure a small vertically oriented eddy rides the mantle stream uplifting the K, holding it in place. This generates northward flow which joins the northern mantle outflow as evidenced by the curvilinear accretion in north Sulawesi. To the east, downwelling micro-eddies in the North Banda Sea, trap the Sulawesi micro-continent continually cycling crustal fragments through the K as it is pushed eastward by the mantle stream following the central vortex. It appears Sulawesi is actually growing northward and the upper limb looks like it will break off and be transported northward on an outflowing mantle stream. The smaller K-structure to the north is trapped by the same processes from a deeper level incoming mantle stream from the Pacific. The K-structures and back arc basins meander along a vortex street trapped in the nebulous surface patterns of the upper level northern outflow mantle stream opposing the incoming mid-level Eurasian and Pacific mantle streams.

9. Pacific Ocean Basin

Several types of tectonics are at play here. Relative motion of the lithosphere, or plate; that is, relative to the deep upper mantle, is shown by the spreading phenomena such as fracture zone azimuths and the ocean floor fabric. Absolute motion of the lithosphere; that is, relative to the surface, could be defined by the linear features, such as the seamount/island chains, or megatrends. This means that two sets of azimuths could co-exist on the same lithosphere, and that absolute motion could change randomly, such as the proposed 43 Ma event. Because of these two sets of trends, relative motion cannot be shown for the Pacific Ocean basin by the fracture zones; they will cross each other somewhere in the western basin. It should be obvious by now that, if absolute plate motion occurs anywhere, it must also occur everywhere else at the same time. For the plate tectonic hypothesis requires that Earth maintain a constant diameter, and the plates are solid rock. Push and pull is the name of the game.

The first, most notable part of the equation to be addressed is the lack of a central portion of the spreading center. A distance of some 2000 km from Mexico to Canada is a transform fault. In the constraints that have been dogmatically emplaced, any spreading in that region, assuming there were a spreading ridge, would necessarily be parallel to the azimuth of the transform. In other words, the fracture zones should all be alined to the NNW, just the same as is the azimuth of the San Andreas Fault. This is not the case as can be seen on any of the Pacific basin maps in this book or any other piece of literature. The fracture zone alignments are perpendicular to that azimuth. Therefore, the North Pacific Ocean floor has no means by which to grow and spread, even though the magnetic lineations show that to be the case. There is no “slab push.”

To that end, the active margins of the Pacific basin are discussed in detail. “Slab pull” at the active margin subduction zones, one of the basic tenets of plate tectonics, is a second driving force of that hypothesis. A large amount of horizontal compression and overthrusting always exists in these settings. All of the intervening oceanic crust is consumed, followed by the abandonment of the subduction zone. The shallow sea in between is filled with sediments, and the mountains begin to rise. These sediment deposits are folded, and igneous rocks are intruded into them.

A passive margin (the shelf, slope, and rise from the plain) is non-volcanic or orogenic. The sediments in the shelf area show abundant evidence of shallow-water deposition and, characteristically, prograde seaward; that is, they get thicker. In the plate tectonic hypothesis, continental rifting precedes passive margin subsidence. It begins with graben formation followed by volcanic activity, such as basaltic dike and sill intrusion. A passive margin can become an active margin when subduction begins along it or because an ocean between it and a land, beneath which the sea floor is subducting, closes.

To show the lack of continuity along the slab front, a sampling of the variable depths includes: The Aleutian Trench ranges from 5800--6900-m, Kuril, 7000 m; Japan, 9000 m; Kashima guyot to the next seamount (Izu), 8600 m; that seamount to Uyeda Ridge (Bonin), 9000 m; Fryer Guyot to another seamount (Mariana), 7000 m; that seamount to delCano Seamount (MAriana), 8200 m; south of Serrao (southern Mariana/ Challenger Deep), 9600 m; Manus, 6000 m; Kilnailau, 4500 m; New Britain, 8000 m; Tonga, 6500 m; Kermadec, 7500 m; north Peru, 7000 m; and Chile, 5300 m.

Seismic stratigraphy has been used in this book. Appendix C shows the interpreted multi-channel seismic profiles for many of the Pacific trench regions. It is to that Appendix that we now direct the discussion. Although data in the area of the Aleutian Trench are scarce, strong signs in that region point toward no subduction. This assumption is based on seismic profiles presented herein by many points: (1) the absence of decollements, which is better interpreted as unconformity or minor thrust triggered by the sediments overloading in the upper and lower slope, (2) flat-lying undisturbed Paleozoic? to Cenozoic trench fills which indicate long lasting stability of the region, (3) landward sediment progradation, which inplies that a seaward landmass existed during the Paleozoic? to Mesozoic, (4) the predominance of vertical block structures in the "oceanic" crust, and (5) the earthquake foci are concentrated under the middle slope (ridge-terrace border) and their absence under the trench.

Further south, the Japan Trench cusp presents the following scenario based on seismic surveys and DSDP Legs 56 and 57, none of which reached acoustic basement. The plate-tectonic history started with: (1) Cretaceous to Oligocene: deposition, landward tilting, and repetition of section by tectonism. (2) Oligocene to Miocene: intrusion by calc-alkalic magmas, erosion, and submergence relative to sealevel. There was deposition of a transgressive sequence of beds. (3) Late Miocene: most of the Oyashio ancient landmass was at sealevel, the seaward edge being bowed down during subsidence. There was an increase in explosive volcanism. An accretionary prism was added as part of the subduction process. (4) Late Miocene to present: subduction accretion above the accreted prism, but the accreted subduction complex is confined to a small area beneath the lower trench slope. Compression stress affects only the frontal part of the upper plate.

Using additional data, a reinterpretation shows otherwise. In the words of Vasil'yev and others (1986): "(1) Paleozoic-Early Cretaceous: sedimentation, alternating with periods of volcanism, and plutonism. At the end of this stage, elevation and erosion. (2) Late Jurassic-Early Cretaceous: areal outpouring of pillow lavas in relatively shallow-water environments and intrusion into the pile of hypabyssal bodies with basic composition. (3) Paleogene: uplift and fracturing. (4) Miocene: slow subsidence, sporadic accumulation of tuffaceous and organic-clay sediments with further terrestrial pyroclastic volcanism in places. (5) Pliocene: compression, and formation of thrusts in the basement of the present arc-trench system. (6) Late Pliocene-Pleistocene: sharp subsidence with formation of faults and tectonic scarps in the trench slope as a result of extension. Formation of trench in its present form and commencement of deposition of its currently undisturbed sediments." Placed within the constraints of the surge-tectonic hypothesis, this explanation is more appropriate for the actual data.

Again in the NW Pacific a similar story is repeated. The deep ocean dredge hauls have revealed that an extensive area in that region is underlain by Precambrian mafic, metamorphic, and granitic rocks, such as quartzite clasts which are overlain by Mesozoic and Cenozoic basalts and sedimentary layers. In the Nankai Trough, thick Paleozoic sedimentary layers appear to cover the Precambrian rocks. The sedimentological, seismic stratigraphical, and paleogeographic studies testify that the wide area of the NW Pacific had been land until the Jurassic and supplied sediments into the paleo seas.

Jumping across the basin, the seismic stratigraphy around Central America is no less compelling, giving a fairly substantial sampling of the circum-Pacific "subduction" zones. The seismic interpretation here is supported by land geology and DSDP drillings. The plate-tectonicists assumed subduction at the landward foot of the trench slope, which is absent in Choi's interpretation. The collapse of the axial area, a result of vertical tectonics, is emphasized. The plate model ignores the unconformity at the surface of the horst and labeled it as a landward dipping reflector. Internal reflectors and unconformities in the horst block and clearly visible layering at the bottom section of the horst were again ignored.

The subduction zone convergent-type margin (Types 1-3 above), or Benioff zone, must follow the Navier-Coulomb maximum shear-stress theory. In a cooling Earth, or the tectogenetic phase of the geotectonic cycle, the lithosphere must dip at an angle of less than 45° because it is in compression. The strictosphere layer must bend at an angle greater than 45° because it is in tension. The two are mutually exclusive in that they do not form a continuous slab and might be offset as much as 100--200-km by the asthenosphere. Cross sections of earthquakes show the two different dip angles and segmentation for the New Hebrides, the Mariana, the Tonga-Kermadec, and the Kuril trenches. In all cases this is the actual fact. Benioff discovered this phenomenon even as he was discovering the Benioff zones themselves. This means that the subducting slab has been known to be decoupled at depth since the phenomenon was first noted. As has been pointed out herein, no continuously dipping plate exists in nature. In an ongoing study of inter-lithosphere deformation at the NW Pacific convergent margins, the hiatus in earthquakes between 150 and 300 km was reconfirmed. Meyerhoff worked with the Navier-Coulomb maximum shear stress theory to prove exactly what Benioff discovered. Thrust-fault fractures developing under compression should dip at 30°. Fractures developing under tension should dip at 60°. There should be no fracturing in between, and there is not. The asthenosphere lies in the range of 150 to 300 km, exactly where few earthquakes have occurred. The "subducting" slab does not keep going down; it veers into the landward, overriding lithosphere as has been repeatedly shown.

As noted, deep earthquakes appear only where seaward features are at the active margin and appear nowhere else. In actuality, if the SW Pacific active margins were removed from the table, negligible earthquake activity below 250 km would exist globally. This means that only one active margin fits the definition of subduction zones from a seismologist's standpoint. So little is known about the bathymetry in that region as to not be able to correlate those deeper earthquakes with subducting aseismic, buoyant highs other than Bougainville Guyot, the nearby re-entrants, and the Louisville Ridge. They are all juxtaposed with the deep earthquakes, and this corroborates previous studies for the NW Pacific convergent margins.

Taking the compression idea one step further, why do the deep earthquakes only exist in the nine discrete locations? The clustered earthquakes all lie below clusters of volcanoes on the active arc. Two possibilities present themselves: either (1) the approaching seamount causes the constriction or (2) the pressure release at the constriction causes the seamount. In explanation, where the constriction occurs, pressure is exerted equally in all directions at the depth of the magma pool. The pressure upward causes a rupture in the lithosphere, which allows for the free passage of magma to the surface. This explains the excess surface earthquake activity and the excess magmatism which builds the increased amount of volcanoes. The downward pressure of the magma at the constriction causes excess pressure at the asthenosphere/upper mantle boundary. The central portion of this geosphere is more fluid, so earthquakes do not occur. In addition, the upper mantle is not as elastic as the asthenosphere, so it passes the excess pressure downward. This excess pressure appears in the data as the deep earthquakes and explains the phenomena of the clustered deep earthquakes and the overlying clustered volcanoes. Lateral pressure exerted at the constriction causes fractures in already weakened lithosphere, a lithosphere that is undergoing excess stress transfer from the aseismic, buoyant high. As the fracture widens, it appears on the surface as a canyon, and it is the path of the breakout surge channel. Thus, the entire microcosm is easily explained by the surge-tectonic hypothesis.

The crust has different ages across the trench. The magnetics measurement and the DSDP and ODP drillsites, which have never reached the real basement of the oceanic crust or had a successful transect across an entire active margin, can neither confirm nor deny any of that theory. Nobody has drilled, cored, or dredged Layers II or III. Nobody has recovered eclogite or spinel. Those phase changes are laboratory experiments. Seismic reflection profiles are our only means of study for these layers.

Segmentation is addressed. The trenches have always been shown as being continuous from start to finish and generally the same depth for each trench respectively. Rather than rely on the past, let us look at the facts. Trench segmentation may show many variables: (1) different types of volcanism may exist across these boundaries, (2) different types of volcanic products may also occur, (3) the separate segments may move independently of each other, and (4) extremely virulent or unusual volcanoes may exist in these boundaries. By using the variable trench depths, this paradigm may show segmentation where none exists bathymetrically.

Where the trenches are segmented, obduction has been the historical explanation. This is also the site of most of the seismic activity and the clustered volcanoes. Where Ranneft demonstrated his ideas, presumably based on land-based transverse faults and oil-field data, the Navy's bathymetry has substantially verified those ideas. The trench segments are the sites of breakout channels in the surge tectonic hypothesis.

Ranneft has already shown many instances of trench segmentation and oblique faulting all over the Pacific Basin. He says that the "hinges" may be related to structural, morphological, or movement aspects of the under-thrusting lithosphere. Now, in the Aleutian region, the same phenomenon is present. The fold axes, which occur as a first-order deformation, are 030°, 060°, and 080° for the entire length. The 060° axis is perpendicular to the convergence angle of oceanic lithosphere. Thrust faults ranging from 055° to 095° can be mapped upslope from the anticlinal folds as the second-order compression deformation. Further up the inner trench wall oblique faults occur. West of 160°W they trend 280°. East of 160°W they trend 020°. A 60 Ma old transverse fault transects the two which passes through an aseismic zone coincidental to Unimak Island. No aseismic gap exists at 160°W, although the location of the old transverse fault is fortuitous in its proximity to the clustered volcanoes across the trench from Derickson Seamount. Subduction has been definitively shown not to have occurred at the margin. The Aleutian "trench" is a strike-slip zone.

The term "subduction zone" is not used because that tectonic event has already been proven to be a non-event. No trench exists off Cascadia (Figure 51). The Aleutian and Vityaz Trenches (Figure 41) are more of a strike-slip zone. Precambrian basement underlies all of the trenches in the Pacific Ocean basin. Deep earthquakes have been theorized to outline the Benioff, or subduction, zone. Benioff showed that the zone was not a continuous feature. Others have also repeated the same. The idea of lithosphere continuity is illusory in that it was introduced by a cartographic sweep of the pen. For the practicing neophyte, this is the fabled cartographic license. All of the earthquakes in the NGDC data base were plotted. As has been shown herein, deep earthquakes occur in only nine spots worldwide, and that is on the landward side of the trenches. Earthquake activity has been proposed by many authors who espouse the plate-tectonic hypothesis to cease where large, buoyant, aseismic features are in the subduction pattern. Quite the opposite has been shown to be the case. In fact, the earthquakes appear to swarm landward of where large, buoyant, aseismic highs are in the trench.

Now there is no more need for **ad hoc** explanation, plate direction changes, or any other eccentricities involved in the explanation of the tectonics of the ocean basins. A mere planetary wobble, and the surge channels continue to flow eastward and leave these marks of their passing. During periods of tectogenesis the roofs collapse to create the typical fracture zone morphology of parallel ridges and valleys, or depressed features. During periods of taphrogenesis the linear seamount chains and oceanic ridges, or raised features, are created. We have seen evidence of this on at least four different azimuths in each of the three large ocean basins. Modern literature is full of ideas about a 54° wobble. The surge channel lineaments rotate through about 120°. All of the active surge channels in the Pacific are on the same azimuth as the Hawaiian chain. This fact leads one to believe that a major adjustment in Earth's rotational axis occurred before that time, causing the constantly eastward-flowing surge channel activity to leave a differently oriented imprint on the ocean floor. The intersection here, commonly called the Hawaiian-Emperor elbow, has been repeatedly used to show plate flow directional change at 43-Ma. However, from a surge tectonic standpoint, this is not necessarily true. A Chandler wobble or any other wobble to Earth's axis changed the direction of flow.

Discrete examples from the text are provided as an explanation of the regional tectonic events, starting with the Obruchev Rise. Bathymetry-wise, the rise is not of a size to elicit excitation. The geomorphology part of the equation is the trench infilling, which is almost nil, and the forearc canyons caused by inter-lithosphere stress. There must be another explanation, and there is. The Obruchev Rise is obviously not responsible for the cusp as has been speculated for the cusps at the Caroline Ridge and the Ogasawara Plateau. The juncture of the Kuril and Aleutian Trenches is 166 km north of the Rise. In surge tectonics, the cusp is always the cause of seaward seamounts, not the reverse as has been proposed and repeatedly shown. Breakout channels occur where two cusps lie. The age of the cusp determines the age of the breakout channel.

The active (arc) margin channel lies about 150-km landward of the trench axis. The active arc, a small scale analog to the greater trunk channel, initially operates independently of the subduction process and is actually an MOR by definition. They are also called volcano, or active, arcs. They are the supply line for the breakout channels, features which figure prominently in a discussion of the western Pacific Ocean basin. Breakout channels are long, linear-to-curvilinear channels that typically underlie linear island and seamount chains. They grow eastward, generally showing a randomly aged sequence, getting younger to the east. The present scars on the ocean floor allow one to interpret the location of the poles-of-rotation when the breakout channels formed. The fact that many of these channels are old has been demonstrated by the Precambrian material lying seaward of many of the North Pacific trenches.

The next two features are breakout channels that lie in the Izu-Bonin cusp. Uyeda Ridge, about 56-km north of Broken-Top Guyot, is definitely fracture controlled on a strike of N71°E, or WSW--ENE. The fracture trend can be followed for 410-km, and the ridge itself is 157-km long. The sinuosity is readily apparent, and this fracture ridge has now been found to be the westernmost, oldest extreme of the Chinook Trough fracture zone. The feature, crossing the trench, continues to the active arc, or the active margin channel.

Events at the Izu-Bonin Trench cusp (Figure 22a) are explained in the newer working hypothesis. The WNW--ESE alignment of the Michelson Ridge is a physical impossibility within the plate-tectonics framework. The problem has existed since this feature was introduced that no spreading ridge existed anywhere near, nor on the proper strike, to produce the Michelson Ridge. Now there is a reason. The appearance of obduction/offscraping/over-thrusting, such as the Ogasawara Paleoland, may be achieved by the development of a breakout surge channel that is still active and has not yet collapsed. The end result gives the same material on each side of the trench. The fact that most of the dredge hauls contained igneous rocks, including rounded cobbles, matches those samples taken from atop Smoot Guyot further to the east on the same Michelson Ridge. Unfortunately, previous studies did not date the igneous rock samples, which have been a big help in establishing the age of the surge channel. Because the Ogasawara Plateau is nearer the source, it is shoaler than the other features. From the geomorphology, the Michelson Ridge guyot tops get deeper and narrower the further they are to the east. By the time the source magma has reached that distance from the active margin channel, about 555-km, it has been reduced in activity commensurate to that distance by producing the smaller Pollux Guyot. Diagrams of this feature on the cover show the magma flow pattern to be easterly and outwardly.

The same scenario is repeated to the south of the Ogasawara Plateau at the trench juncture of the Magellan Seamounts (Figure 22c). The seaward seamounts appear to extend onto the forearc as two legs of a "Z" formation, called the Lapulapu Ridge. Extension of the "Z" legs to the trench axis shows both of them to be offset from their respective counterparts on the seaward lithosphere, delCano and Quesada. The seaward lithosphere, remember, is converging obliquely. This means that the Magellan Seamounts have moved north about 30-km since the formation of the Lapulapu Ridge. One has only to look at Quesada and the northern rubble to see the south trending hook. The earthquakes show subduction to the west on the south and subduction to the east on the northern portion. Realizing the physical impossibility of such a statement, there must be another solution to actual occurrences for this region.

Furthermore, the earthquake foci distribution in the profiles of Kuril and Japan Trenches show that the shallow earthquakes are densest in the middle continental slope, some 70 to 140 km landward from the trenches, but not at the trenches where the Pacific plate is presumed to start subducting at the speed of 10 to 10.5 cm/year. The similar fact is seen in many of the Pacific trenches such as the Aleutian and Peru-Chile Trenches mentioned elsewhere.

In a landmark study of events in Southeast Asia by Art Meyerhoff, the timing of all of the surge channels flowing out of the landmass into the Philippine Basin have been delineated. The largest feeder channel for this region comes out of Sulawesi and turned north behind the Philippine Trench. This channel has been in existence for 610 Ma, or since the Sinian. Another fork welled up behind the southern Philippine Islands, a central fork passed through a gap in the trenches at 10°N, and the northern fork passed NE behind the Ryukyu Trench during the Carboniferous-to-Early Permian (362-280 Ma). The status quo remained here until the Late Jurassic (150-145 Ma). From then until the Middle Eocene (40 Ma) the surge channel formed breakouts and poured through gaps in the Benioff zones out into the Philippine Sea region. During the Cretaceous (145-65 Ma) the channels found their way south through gaps in the Benioff zone. One of these was a SE-flowing fork now called the Manila-Negros-Sulu Trench complex, the eastward pointing northern part of the engulfment zone at the Philippine Islands.

E-W striking magnetic anomalies exist in the Philippine Sea basin, which are interpreted as breakout channels from the Ryukyu-Taiwan-Philippine region (Figure 49). The anomalies range from Early--Late Cretaceous. These could be older. They are on strike with anomalies in the region that are believed to be as old as Carboniferous.

Magma in the breakout channels coming off the active margin channel from the Japan Trench-Nankai Trough intersection seeped through gaps in the Mariana, Bonin, Izu, and Japan trenches to cause the Mid-Cretaceous (120-110 Ma) outpouring of volcanoes in the western Pacific basin. The bathymetric expressions of that voluminous outpouring are: (1) the Uyeda Ridge, which continues east as the Chinook Fracture Zone, and its partner as a K-structure, the Michelson Ridge, (2) the Dutton Ridge, which continues northeasterly as the Mendocino Fracture Zone, (3) the Magellan Seamounts, which continue SE as the Marshall-Gilbert trend, and (4) the Caroline Ridge on the south side of the Mariana Trench.

These anomalous ridge trends may be caused by increased amounts of magma outflow along preexisting faults. Similar magmatic outflows have been witnessed subaerially on the island of Hawaii. Given points of weakness and rupture of the lithosphere along a tectonically active area, there will be magma rising to the surface. An evolutionary scenario is suggested in which the Chinook Fracture Zone became active before the emplacement of either the Hess or Shatskiy Rise. The activity in the oblique features makes it possible to explain the presence and prevalence of the 040°-striking, oblique ridges in the 060°-trending Chinook Fracture Zone. These ridges have necessarily formed at the time when differential lithosphere motion was taking place. That motion provided the mechanism for the obliqueness of the ridge trends and bathymetric signature not associated with active tears. The rises formed after the fracture's propagation eastward; that is, after tectogenesis when the planet was on another pole-of-rotation.

Isostatic compensation influenced by the rises is discounted as an agent of formation. Instead, it is merely one of propagation. The correct route this portion of the north-central Pacific Basin has already been demonstrated for the Mendocino, Pioneer, and Murray fracture zones and, in fact, the eastern Chinook Trough itself; both avenues are correct, and they merge in the central Shatskiy Rise. While the age of the rises remain somewhat in question, the ages of the fractures on the WSW--ENE azimuth are definitely in question.

A certain amount of compression is happening at this point at witnessed by the configuration of the aptly-named Snake Ridge (Figure 17). The question remains as to whether that ridge formed first as a linear fracture-zone ridge and became distorted through the compressive forces during a previous geotectonic cycle or whether the ridge formed **in situ**.

These fracture zones/swarms were formed when the planet was on a pole of rotation other than the present pole. Their present azimuth is generally WSW--ENE. Because the primary constituents in the geomorphology are fracture valleys, the features must of necessity and by definition be older than the active surge channels. The older features are subject to reactivation, and many of them do display cross-trend seamounts and ridges within the near boundaries of those features. The younger expressions of surge channel activity are the ridges and seamount chains as has been previously noted, and their present azimuth is generally WNW--ESE.

GEOSAT data (Figure 16) and mid-plate earthquake seismicity for the central Pacific Ocean basin allow us to continue the easterly-flowing trend from the Golden Dragon's lair, 180°, to 80°E longitude. The advantage to using the high-pass filtered GEOSAT data here is that the surge channel is only partially manifested surficially because it is still active. This surge channel has been active for the past 43 Ma.

In a surge-tectonic setting, the East Pacific Rise and the San Andreas Fault, which is an almost extinct section of this trunk channel, are one in the same. The feeder channel coming from the north, which is represented by the Gorda, Endeavor, and Juan de Fuca ridges, intersect the older surge channel of the Mendocino Fracture Zone to form the magma flood, called the Columbia River Basalts. This geostream continues to the SSE under the San Andreas Fault to go offshore as the northern portion of the East Pacific Rise. Where the East Pacific Rise bows eastward the most in the bathymetry, the location of what they term as a triple junction, is the site of a feeder channel going east. Rather than being defined as the place where the Pacific, Cocos, and Nazca plates meet, it is the site of the beginning of that feeder channel that will ultimately form the eastward bulge of the outer Caribbean Island arc. In the vernacular, this region is called the Galapagos Rift, and part of it is a vortex. called the De Steiguer Deep.

The next branch of the active trans-Pacific surge channel begins at the Fossa Magna region in Japan. It crosses the Japan-Izu cusp to become the Geisha Guyots, the Hawaiian Ridge, the eastern portion of the Mid-Pacific Mountains, and the acutely Murray Fracture Zone-crossing Moonless Mountains, Erben Guyot, and Fieberling Guyot. From the GEOSAT diagram the northern segment into Clarion, the southern segment out of Clipperton, and an intersection with the East Pacific Rise trunk channel at the De Steiguer Deep, which has something of the appearance of a vortex, are included. The vortex lies at 2°30'N latitude and 95°30'W longitude. From there, after its roundabout, it continues as the southern segment of the Galapagos "propagating rift" to the Cocos Ridge.

Please return to the western Pacific basin once again. For the southern portion of the Philippine Sea region, the magnetic anomalies strike N-S behind the trench system. The anomalies of eastern Indonesia strike ENE. Those of the Bismark and Solomon Seas strike northeasterly. All converge on the Caroline Basin and continue beneath the Ontong-Java Plateau (Figure 39), which is the site of the largest cusp in the world. Made up of the Yap and Palau trenches on the north, Sulawesi on the west, and Irian Jaya and Papau New Guinea, and the New Britain, South Solomon, and Vityaz trenches on the south, this is the type of feature where one would necessarily go to look for a breakout surge channel. Because the splayed surge channels were in existence for the 410 Ma preceding the end of the Ordovician, the western Pacific Benioff zones must have necessarily already been in place. In fact, sampling has given at least a Paleozoic-aged basement to the Philippine Islands, Proterozoic granitic clasts in western New Guinea (1,250 Ma) and NW Irian Jaya, 1,700-1,500 Ma on western Malay Peninsula. The Ontong-Java Plateau is dated at 129-83 Ma. Flow is still taking place beneath this region as witnessed by the earthquake and volcanic activity. As the flow increases, the surge channel revolves counterclockwise and upwells. The GEOSAT diagram (Figure 16) indicates a massive surge channel pouring out to the east from that region.

The Yap Trench and the West Mariana Ridge/ South Honshu Ridge are perfectly aligned, and this is the inactive surge channel that formed to the east of the Palau Trench/Palau-Kyushu Ridge inactive surge channel. The Caroline Ridge is the active breakout surge channel, a part of the active margin channel flowing behind the NW Pacific Benioff zones, called the East Mariana Ridge. Using the surge tectonic hypothesis about active margin and breakout channels, that channel has apparently made several attempts to break out along the segmented trenches north of here, such as at the Izu-Bonin cusp and at the Magellan Seamounts. At the Mariana-Yap cusp nature has unleashed it's pent-up forces, forces which are ever trying to flow easterly. The prior existence of the Sorol Trough and the West Caroline Trough to the south of that are keeping the flow path in check. Ulithi Atoll is only the first of many subaerial excressences due to excess magma flow. The active breakout surge channel joins the ESE-flowing trunk channel coming from the Philippine Trench and New Guinea.

A segment of the surge channel extends through New Guinea and splits at the New Britain (north) and New Hebrides (south) trenches. The north fork continues co-parallel to the Caroline Ridge channel and passes beneath the Ontong-Java Plateau. The two surge channels cross the Pacific Basin through the Samoan Islands, the Manihiki Plateau, still continue as a double trace through the Tuamotu Archipelago, and to the Easter Island vortex. The south fork continues flowing southeast, crossing the domain of the mythical Shellback until it converges with a channel coming northeast along the Pacific-Antarctic Ridge at the Kermadec and Tonga trenches. A giant-sized vortex, the Fiji Plateau (Figure 48), is formed at that juncture. This is a region of upwelling. The geostream continues across the Pacific basin on a ESE direction to intersect the Easter Island vortex on the East Pacific Rise at 24°S latitude and 113°W longitude. That point is the site of major events.

The Easter Island downwelling vortex is the eastern site of one of the controls of the oceanographic phenomenon known as El Nino. Part of the geostream continues eastward as the Sala y Gomez, Nazca, and Cocos ridges, trending ENE to South America. The major portion coming out of the roundabout route at the Easter Island vortex continues ESE as the Chile Rise. A portion travels along the East Pacific Rise until it veers off to the east as the Tehuantepec Ridge, a feeder channel.

That is the general outline of the formation of the Pacific basin. As Earth wobbled on its axis, the pole-of-rotation changed. The compression-formed fractures also underwent realignment. Just as the present pole is where it is, so is the formation of the breakout surge channels and the fracture zones indicative of the location of those poles.

10. Tectonics around South America

DeMets and his working group have proposed that the Atlantic basin is moving to the NW. The Caribbean and Scotia regions are moving eastward. Stress is being produced. According to Mary Lou Zoback and Kevin Burke (1993) the stress on the south is relieved by normal faulting, and the stress release on the north is strike-slip. DeMets and others also show the "plate boundary" between the North and South American "plates" to intersect at the SE limit of the Puerto Rican Trench at 17°N latitude. There is neither bathymetric nor geophysical evidence for this arbitrary placement of the boundary. In fact, the boundary has been placed at 15°N latitude and 19°N latitude.

Evolutionary events leading to the current geomorphology found in this region seem very comprehensive at first glance. W.D. Cunningham and others (1995) think that South America and Antarctica have been separating since 84 Ma, with South America moving westward more rapidly than Antarctica. This has caused 1320 km of E--W left-lateral strike-slip displacement and 490 km of N--S divergence, resulting in the opening of the Drake Passage. An angle change during the global Eocene plate reorganization (Hawaiian-Emperor elbow, rise of the Himalayan Mountains, etc.) led to the onset of ocean floor spreading in the Scotia Sea at 30 Ma. This is presumably also the onset of subduction, creating the South Sandwich Trench during this time. Subduction retreat along the South Scotia Ridge and South Sandwich arc, combined with backarc spreading contributed to the width of the gap between the continents. The South Scotia Ridge is the trunk channel, and the Drake Passage is marked by the Shackelton Fracture Zone.

As noted by the German-Russian study, none of this is true. The separation is caused by lithosphere stretching of the Precambrian bridge.

During the Precambrian the active surge channels passed through the South American continent (Figure 50). In a landmark study, Choi has identified erroneous data and corrected same. Carefully traced reflectors which are considered geologically significant were used for the study.

Additionally, on the crustal profile across the Peru Trench and the earthquake foci (see Appendices B and C), the earthquake hypocenters are densest under the middle continental slope between the trench and the coast (at around 80 to 150 km east of the Peru Trench) with a depth range of 25 to 70 km from the sea level. Several very shallow earthquakes (less than 10 km from the sea floor) are located in this area too. Whereas, only three hypocenters are scattered under the Trench at about 25 to 30 km in depth in the mantle. In addition, no earthquakes have been recorded along the alleged megathrust or the upper surface of the underplating oceanic plate (the 6 km/sec layer) as illustrated by others. The seismically active Wadati-Benioff zone is not located at the trench, but about 80 to 150 km landward from the trench. The situation is similar to the Aleutian Trench as noted by Murdock in 1998, Japan Trench, and Kuril-Kamchatka Trench. The available evidence, 1) presumed paleolands, 2) continuation of continental Precambrian structures into the ocean floor, 3) new seismic profile interpretations as shown by Choi and 4) the above mentioned Benioff zone and crustal structure in the Peru Trench area, would indicate that plate subduction has not taken place along the Peru and other trenches in the Pacific.

If no subduction is taking place, and if the basement rocks are Precambrian in age, then the surge channel activity at that time continued through the continental craton. This occurred in at least three different places. Essentially, the great trans-Pacific geostream split into three segments at the East Pacific Rise. One went ashore as the Chile Rise at the Taitao Peninsula to continue under the Paleozoic Deseado Massif to pass into the South Atlantic basin under the Falkland Islands. A second surge channel continues into the continental craton from the Nazca Ridge under the Proterozoic Brazilian Shield. A third surge channel comes in from the Galapagos Islands and the Carnegie Ridge to pass under the Proterozoic Guyana Shield. All of these are extinct surge channels.

Choi has extensively researched western South America and its relation to the subduction process (**Active Margin Tectonics**, 2001). The Carnegie Ridge goes into the South American continent at least through the Guyana Shield. In a study of the seismic data, major Precambrian (Proterozoic) structural trends in the South American continent clearly continue into the SE Pacific. Published geologic and tectonic maps show that major Precambrian (particularly Proterozoic) structural trends recognized in the Precambrian shield connect to the main bathymetric features of the SE Pacific Ocean basin. That flow is instead splitting and passing around the continental extremes. this explains why the Caribbean and Scotia arcs appear as they do. The two different passages are dealt with as (1) the Caribbean eastward bulge and the (2) Scotia eastward bulge.

The Caribbean bulge is a northern branch of the surge channel which went ashore in northern South America as the Carnegie Ridge has split off to become the Cocos Ridge. That crosses Guatemala in the form of the Polochic Fault and Motagua Fault. It is joined by Russo and Silver's bifurcating magma flow around South America. The seismic interpretation, supported by land geology and DSDP drillings, largely contradicts the geologic interpretation made by plate tectonicists in the same area. They assumed the subduction at the landward foot of the trench slope which is absent in our interpretation which emphasized the collapse of the axial area, result of vertical tectonics. The plate model neglected unconformity at the surface of the horst, and labeled it as landward dipping reflectors. Internal reflectors and unconformities in the horst block (Ib to Id) and clearly visible layering at the bottom section of the horst (Ia) were totally neglected. As a whole the plate interpretation lacks in geologic consistency.

Another important fact came to light by our analysis is that the Middle America Trench is formed on or near the axis of the Precambrian basement highs. In all trenches in the Pacific so far examined, it is shown that the trenches are situated on or near the axis of anticlinal structure of the lower crust (at least partly Precambrian in age and composed of mafic and continental rocks) that has subjected to intensive vertical tectonics as well as long lasting subaerial erosion before submerging to the present depth.

With the predicted lack of stress on the northern boundary, no reason exists for an active arc in the Caribbean region. However, many extremely active volcanoes are in this region on the Antilles arc (from many Smithsonian Institution SEAN Bulletins). Mt. Pelee is on Martinique, La Soufriere on St. Vincent, and Kick-'em-Jenny off Granada are the most noteworthy. All are between 12° and 15°N latitude. The active arc is segmented, showing a NE-trending transverse fault north of Guadalupe and La Desiderade Islands, an eastward-trending transverse fault between Martinique and St. Lucia, and a SE-trending transverse fault south of Granada. These are compounded by the effects of the St. Lucia-Barbados Crosswarp and are probably responsible for the intense volcanic activity in the region rather than the subduction process itself. In fact, surprisingly few earthquakes exist.

The Puerto Rican "Trench" is a theoretical subduction zone, and it appears to have been squeezed out between North and South America. There are no large, aseismic, buoyant highs in the trench axis to interfere with subduction. There is no trench on the eastern side of the Caribbean bulge where subduction should be occurring at the juncture with the westerly-moving Atlantic plate. The Puerto Rican Trench cannot be traced any farther SE than 17°N latitude and 59.5°W longitude for a start, so we are left with the distance from there to the South American continent with no trench for subduction. There are many surface earthquakes and no deep, so the trench has probably been overprinted by sediment. The trench continues to the west until it intersects the Oriente Fracture Zone, more-or-less. The earthquakes stop at about 74.5°W longitude. There is no geophysical boundary on the south.

Table V. EARTHQUAKE DATA FOR THE CARIBBEAN REGION

(data from the 1990 NGDC earthquake CD-ROM)

Earthquake depth (km)	Quantity
0-49	3308
50-99	442
100-149	236
150-199	98
200-249	4
250-700	0
unknown	108

At a loss to explain the actual events, we shall retreat under the aegis of the new working hypothesis once again. The three surge channels in the Caribbean region arise off the East Pacific Rise. The Antilles and Venezuela channels enter through the Tehuantepec Ridge, which is as old as the early Paleozoic (about 550 Ma), and the Cocos Ridge (undated). The channels are at least as old as the Carboniferous (362-290 Ma) and very likely date to the late Proterozoic (1,000-610 Ma). The Columbia-Ecuador system enters through a branch of the Galapagos rift, the Panama Fracture Zone. This avenue is known to have been in existence since the Triassic (245-208 Ma). All of the surge channels are older than the forecasted 200 Ma maximum ocean floor age predicted by plate tectonics. All are based on actual rock sample dates from the features in question.

A combination of studies on the Nazca active margin has given the answer to what kind(s) of tectonics are at play here. Different investigators have predicted eastward flow of the lithosphere around the tips of South America, and that is exactly why this region is acting as it is. Realizing that all oceans are antipodal to continents and that contraction accounts for all measurable shortening in mobile belts, Morris and others made the low-velocity zone above the asthenosphere to be very mobile. This was the forerunner of the surge channel concept and is summarized as follows: (1) Extensive Triassic-Middle Jurassic Indosinian orogeny in Asia triggered a massive surge from west to east across the Pacific Basin, and this surge exerted pressure against the South American craton. (2) The pressure generated by the Asian orogeny caused a rupture of the paleo-isthmus that connected North and South America, creating the Nicaragua Rise during the Middle Jurassic. This surge channel and the one south of South America are prominently shown by seismotomography. (3) There were three active surge channels working on this region to create the bivergent foldbelts; the Antilles system on the north, the parallel Venezuela system on the south, and the Columbia-Ecuador system along South America's northwestern coast. The magma surge eastward caused the isthmus to rotate counterclockwise, which produced a series of magmatic arcs in the region of influence. These arcs are the Cuban arc, which shifted north until a collision with the Florida-Bahamas block stopped motion there; the Antilles arc, which collided with the same block during the Eocene; the Aves arc, which moves eastward to create the Lesser Antilles arc during the Middle Eocene; the Paleo-isthmus arc, which collided with the Yucatan block in the west during the Eocene-Oligocene and the Main Antilles arc in the east; the Venezuela arc and the Barbados Ridge extension, which shifted south until its collision with the Guayana shield; and the Middle America-Columbia magmatic arcs that face the Pacific. This was systemmatically destroyed and replaced by the Panama magmatic arc and the Middle America Trench. (4) Repeated orogeny during this time frame, the Cretaceous to the early Tertiary, ended with the Eocene "paroxsymal logjam" in the Caribbean of dead arcs and segments of arcs and the newly created ridges. Wrench faults, tear faults, and vertical tectonics gradually became the norm, and the surges tilted the region to the east.

Several lines of proof have opened for these explanations. The CAYTROUGH (1978) group collected 142 rock samples from the Cayman Trough (Figure 3-18) "spreading center's" inner walls using the deep-submersible, ALVIN. Most of the samples should have been fresh basalt in a plate-tectonics framework. Most of the samples were actually gabbroic in nature, with some gabbroes cut by foliated amphibolites. Gabbro is generally considered to be a Layer 3 rock in an oceanic setting. No basalts were recovered. The other dive area was outside the valley on a bathymetric high. It produced all basalts from pillow lavas. The oceanic lithosphere under this segment is only 250 m deep. Interestingly, a study based on ODP Leg 156 in 1995 has drilled three holes across the accretionary prism at 15°32'N latitude, roughly along the Dominica Fault. While no subduction apparent is in their published seismic trace (see their figure 1, pg. 185), the crew believes they are setting a precedent by leaving borehole seals that will give continuous measurements of pressure and temperature for two years. The results may help in unraveling this enigma to the tectonic theories. Finally, ODP Leg 165 drilled Sites 998-1002 in the Caribbean Sea. Site 1001 reached igneous basement. Arc volcanism was documented along the Cayman Ridge as well throughout the basin for the Eocene and Miocene. Many of the facts led the investigative group to declare the Caribbean "Plate" to be a large igneous province, **a la** the Coffin and Eldholm study (1996). It simply appears to be another of the magma floods.

As for events south of the Caribbean bulge, the Scotia bulge (Figure 5 bottom) appears to be identical, with the active surge channel providing the impetus for the proposed stretching of that continental crust.

Table VI. EARTHQUAKE DATA FOR THE SCOTIA SEA REGION
(data from the 1990 NGDC earthquake CD-ROM)

Earthquake depth (kms)	Quantity
0-49	1276
50-99	159
100-149	138
150-199	29
200-249	1
250-299	1
300-700	0
unknown	36

All of the earthquakes are essentially above the strictosphere in the surge channel/crust range (Table VI). The earthquakes define a region of flow along the southern portion of this plate turning counter-clockwise to the north and northwest. By the point at 33°W longitude, earthquake activity has ceased.

The GEOSAT gives no indication of seamounts at the active margin. The earthquake regime shows a clustering of epicenters on the north and southeast portions of the active arc, suggesting the presence of a surge channel. A GEOSAT diagram shows the existence of old, cold highs and lows on the ocean floor. On the extreme northwest the flat area is the Falkland Islands plateau, and the clean area on the southwest is the South Shephard and the South Orkney Islands. Proceeding through the Drake Passage on the west across the southern portion of the Chile Trench, the South Georgia Ridge to the north is next. This turns south. Fractures off the Atlantic-Antarctic Ridge delineate the southeast portion of the diagram. This leaves the Scotia arc as the easternmost high, and the South Sandwich Trench as the easternmost, seaward low. The southern boundary is a trunk channel passing through from the Pacific Basin.

11. Atlantic Ocean Basin

The discussion continues with the next basin to the east, the Atlantic Ocean basin. Worldwide, the lithosphere moves around Earth in a fashion determined by Earth's rotation and the compression forces due to contraction. However, this has not been the case according to scientific methods of the past half-century. The present explanation is that the North Atlantic is the least understood ocean basin on Earth. First, Bullard followed by Bob Dietz presented rudimentary opening pattern fits that helped explain the original Wegenerian theory of Atlantic opening. Then the oceanographers started collecting concrete data to support that theory. We have seen how these data have been massaged and manipulated. As an example, the World Magnetic Model was used to help define the plate-tectonic hypothesis. That model was constructed at NAVOCEANO using "two million miles of airborne magnetic lines, most collected at an altitude of 25,000 feet having a highly variable accuracy" (Fred Valentine, Magnetics Division). On close inspection very little data has been collected between anomalies 13 and MO, leaving tectonic models with a gap in records, a "Dark Age." In the plate-tectonics paradigm, this is an 82 Ma gap between 36 Ma and 118 Ma.

A plate-tectonic history of continental drift and the opening, closing, and opening of the Atlantic Ocean since the Ordovician begins (Figure 2) with two major continents and the sea called the Iapetus Ocean has been described previously. The segment of the MAR adjacent to Iceland has been used constantly to prove the plate-tectonic hypothesis. The Iceland Plateau had been thought to have been formed as part of the British-Arctic plateau-basalt province which had broken up during the initial rifting of the Atlantic Ocean and sunk below sealevel. Because of Iceland's position on the MAR, investigators decided to have Iceland formed on the ridge-crest by volcanic eruptions. In a later explanation, a hotspot was felt to be the creator. Yet later work by V.V. Beloussov, a past-vice president of the IUGG, and Milanovsky (1977) theorized at least three rifts through Iceland and no extensional tectonics (read ocean floor spreading) because the neovolcanic zone overlay Pliocene stratigraphy. Several scenarios have been offered, among them active ocean floor spreading started between the Norwegian Sea-Greenland Sea-Arctic Ocean in the normally accepted fashion (Bullard fit).

The first question about North Atlantic opening was asked in relation to the Oceanographer Fracture Zone (Figure 10), which was thought not to be on the central Atlantic crustal block. A minimum of 8 poles of rotation for the last 80 Ma of North America/Africa plate spreading, with the older spreading following the usually accepted flow lines, was proposed. These flow lines have been calculated with stage poles or estimated from fracture zone trends. However, the flow lines west of 26°50'W longitude do not agree with mapped fracture zone locations.

As more and more new hypotheses proliferated, the magnetic lineations used to "verify" ocean floor spreading on the Reykjanes Ridge, the portion of the Mid-Atlantic Ridge south of Iceland (Figure 5) were called into question by Bill Agocs and Karoly Kis (1992). At this writing the Iceland Plateau seems to have been continental at one time, at least 700 Ma. Matching rocks and mountain ranges is difficult. The theory of the Wilson Cycle is widely accepted, but it is wrong. The Iceland Plateau was part of a large North Atlantic dam that included Greenland, Iceland, the Faeroes, and Rockall Bank. No “spreading” occurred here yet, or Iceland would have the appearance of the Corner/Cruiser Plateau. The Hayes Fracture Zone (Figure 10) has been shown in it's entirety, and it too had disproven the Bullard fit while showing how the separation of the Corner Seamounts and the Cruiser Guyot occurred. Also, young lavas can hardly overlay older unless there is no spreading.

At the 2005 International Geologic Congress a new map was on display. A crustal profile compiled by the Information Division at the Norwegian Survey of the North Atlantic basin clearly shows a pre-Devonian continental block in the middle of the ocean; that is, the Jan Mayen Ridge (Figure 5 top). Pre-Jurassic continental crust extends seaward off Norway

For the South Atlantic Ocean basin evolution, we return to the plate-tectonic hypothesis for an initial explanation. The formation of the South Atlantic Ocean caused subduction to begin on the western margin of the American plate, beginning about 245 Ma. The process began with the breakup of Gondwanaland, which triggered widespread volcanism. After South America and Africa split apart, a compressional regime evolved. This caused an increased rate of subduction and the deformation of the Andean cordillera. This compression has produced 240 km of crustal shortening, a phenomenon whereby the continental crust is piled higher and deeper while maintaining the same volume. This is not true because of reasons listed before.

As far as the ocean is concerned, being that it is still 70% of Earth's surface, nobody has found any dinosaur bones on the ocean floor. Paleobiogeography remains a moot point as to the prior existence of land bridges. However, enough continental-style and old age rocks have been found in the Atlantic basin from the equator north through Greenland that many scientists are claiming that the Atlantic Ocean has always been the size it is now, and that free passage back and forth between Europe and America was a fact. This is true for the land bridge through Greenland/Iceland as well as the one across the equator. The fossil evidence is the same on both sides of the basin. Some shallow-water fossils have been found in the King's Trough region (Figure 10), a region now lying several thousand meters deep. Between the fossils and the continental rocks, one must pause to reflect on the magnanimity of these observations.

A further problem exists in the GEOSAT lineations. Were the Atlantic basin to open simultaneously because two different plates do not exist there, we would have a mountain range considerably larger even than the Himalayas in the middle of the Sahara Desert. The GEOSAT structural trends, or surge channels, show an WNW--ESE overall pattern in the North Atlantic. They show an overall WSW--ENE pattern in the South Atlantic. Coupled with movement by the proposed Arabian plate, all of this compression must necessarily be taken up by the central African plate. Compression features are mountains, and none exist in that region.

Many lineaments seem to be pouring into the Atlantic basin from North America, the Caribbean Sea, South America, and the Scotia Sea. All of these lineaments align with existing fracture zones; all are at least Precambrian in age. The Atlantic Ocean floor is littered it would seem with rocks in the billion-year-old range and older. Several of these surge channels are followed. Because the fracture traces record the inactive routes of surge channels, the active feeder channel in the northern Atlantic seems to be the Azores-Gibraltar Ridge. The active surge channel for the southern Atlantic basin seems to be the Walvis Ridge, and this is how we get from the Pacific to the Atlantic.

The Cayman Trough (Figure 3-18) lineament continues into the Barracuda Fracture Zone (Figure 10) swarm. This is a major swarm in that it merges on the SE with the lineaments underlying the Guyana Shield and the Brazilian Shield (Figure 50). This swarm crosses the Atlantic basin to go ashore at a Precambrian fault zone in Central Africa, a fault zone which appears to the south of the Sahara Desert. Because this lineament is based solely on fractures, the age is presumed to be that of the rocks at the ends; that is, Precambrian.

The active surge channel for the South Atlantic lies mid-basin. The Brazilian Shield surge channel bifurcates, and a southerly-trending arm is part of a Proterozoic geanticlinal high. It continues onto the ocean floor as the WNW--ESE-trending Santos Plateau at 25°S latitude, crosses through the Rio Grande Rise as a fracture, becomes a 500-km long trough at 5200 m deep, and bifurcates to cross the Mid-Atlantic Ridge in the form of the Cox Fracture Zone at 32°S latitude and the Meteor Fracture Zone at 34°S latitude. It continues northeasterly to Africa through Wust Seamount and blends with the Walvis Ridge (Figure12), an active feeder channel. The Walvis Ridge is an extension of the Tristan da Cunha Fracture Zone at 37°S latitude (Figure 12-13). Geological mapping shows that the Walvis Ridge continues on land as an Archaean high at the Namibia/Angola border. It appears to travel northeasterly until it intersects the Great African Rift Valley, where it offsets Lake Tanganyika and Lake Malawi. The northern fork of the Brazilian surge channel splits off to the north at the Brazilian Seamounts to cross the Mid-Atlantic Ridge as the NE-trending Bathymetrists Seamounts.

The surge channel which crosses the South American Deseado Massif continues under the Falkland Islands, skirts along the Scotia Arc, and disappears, temporarily, at the Atlantic-Indian Ridge.

CONCLUSION

We began with an open mind. After an Introduction to the derivation of the current working hypothesis from the first geomorphology studies, we learned of the newer tools for collecting ocean floor data since 1966. Using bathymetry that had been surveyed/compiled since the inception of the multibeam sonar collector, called SASS, in 1967, large-scale charts of select feature types were shown, features formed by both primary and secondary geomorphological forces. Any available accurate information, such as rock ages, earthquake seismicity, seismic stratigraphy, and satellite altimetry data was included in many of the features discussed herein: midocean ridges, fracture zones, seamounts, plateaus and rises, convergent margins, canyons, seachannels, fans, and abyssal plains.

At the end of the introductory truths, all of which were presented on locator maps of the discrete basins (Figures 3, 8, 9, 12, 13, 15, 22, 26, 30, 33, 49, and 57), the collection of the bathymetry and rock ages began the ultimate goal of the ocean floor geomorphology, with the addition of the tectonics. All the while the very definition of geomorphology was kept in sight. To that end, the very essence of the plate tectonic hypothesis, continental drift, was immediately disproven by the chapter on Gondwanaland. With no spreading between any of the continents proposed to comprise that feature, we were allowed to explore further possibilities based on real data.

Along the way we lost the ability of the fracture zones to point the direction of seafloor spreading (Figures 8, 9, 10,11, 14, 16, 36, 42, 43, 48, 51, and 59). The hotspot/magma plume concept has fallen by the wayside. Most plates do not have the requisite boundaries. As an example of that, the African (Figure 12) and Antarctic plates are almost totally surrounded by the midocean ridge spreading centers. And, the spreading centers, rather than being comprised of zero-aged basalts, are at least one million years old and found to be lying atop crust considered to be continental in origin.

The seamount numbers are vastly less than those that have been proposed (Figure 10). The chains lie at the ends of the fracture zones, or within them, making them part of a feature called the megatrend. Most of the salient bathymetry, especially that of the western Pacific basin, has the appearance of eastward flowing structures, presumably based upon Earth's rotation. That being the case, we can easily detect past polar wander as witnessed by the up to four different sets of lineaments imprinted on the ocean floor. These lineaments, or megatrends, also pass through the continents.

With deep earthquakes occurring in only nine spots worldwide, a re-investigation of the descending slab/subduction zone became necessary. The earthquake hypocenters do not describe a continuously descending slab (Appendix B). In fact, in some cases they delineate a strike-slip zone by the fact that they are stacked, such as the Aleutian and Vityaz trenches. The re-interpreted multi-channel seismic traces, where ground-truthing is available, show progradation towards the continents in many cases while all have basements that are Precambrian and continental in origin (Appendix C).

Last is the geometry problem, which cannot be theorized away.

More probably exist, though that is enough to cast the shadow of doubt across the plate tectonic hypothesis and is, in fact, enough to warrant a total paradigm shift. But, rejection without offering an alternative is useless. We have the orthogonally intersecting megatrends underlain by what I call hotlines, real ocean floor ages, Earth rotation, polar wander, and gravity of course. The large, igneous provinces lie atop the intersections. Many of the megatrends seem to pass through the continents; others skirt them flowing eastward. The same types of fossils are found everywhere, and India has never been anywhere but up-and-down. Horizontal movement has been shown not to have existed, so the continental drift idea is passe. Instead, the Atlantic basin is home to several trans-basin land bridges, as was the same from Antarctica.

From the annals of tectonism I have gleaned several of the more interesting possibilities which seem to fit the actual data. For the older lineaments, wrench tectonics is an aid in describing the major lineaments from the Precambrian to the Jurassic. The different megatrend orientations are explained by the different poles-of-rotation if we accept that Earth's rotation is a major player. In order for that to happen, some form of magma flow has to exist beneath Earth's surface simply because of the fluid dynamics aspect expressed on Earth's surface. These are the hotlines.

This brings us to the present. With the exception of the polar wander, surge tectonics fits the hotline concept to a "T." Surge tectonics has hotlines in the form of surge channels. As the channels empty, they collapse on themselves due to gravity and leave a scar, in this case the fracture zones (Figures 11, 14, and 16). In fact, that is the basis of the hypothesis. Leaky fractures are the norm. This is the only hypothesis that recognizes/explains the vortex structures. The trenches are a form of contraction cracks as Earth continues to cool, with an accompanying reduction in waist size. Nothing exists in the real data that this hypothesis does not describe.

Apparently, just as beauty is in the eyes of the beholder, so is tectonics. Nothing can really be proven but the ephemeral appearance of the surface structures and the age and composition of the rocks. The intent is for the interested reader to assimilate all the bathymetry and rock ages and compositions, any of the various hypotheses felt to be germane, and conceive of a more robust working hypothesis of Earth dynamics. The one we have now does not work.

We still cannot predict when or where earthquakes will occur, we are not much closer to being able to predict volcanic eruption timing, and much of the world's ocean floors remain to be mapped. We know more about the surface of Mars and Venus. Hopefully, this book is a synopsis of the current data bases and a step in the right direction.

SELECTED READING

-Active Margin Geomorphology, (X-libris Corp. Philadelphia USA), 164 p, 2001, N.C. Smoot, D.R. Choi, and M.I. Bhat.

-Bulletin of the Global Volcanic Network (National Museum of Natural History, Washington DC), various.

-Critical Aspects of Plate Tectonics Theory, 1990, V. Beloussov, H. G. Bevis, K. A W. Crook, D. Monopolis, H. G. Owen, S. K. Runcorn, C. Scalera, W. F. Tanner, S. T. Tassos, H. Termier, U. Walzer and S. S. Augustithis (eds).

-The Geology of North America, (Geological Society of America, Boulder [D-NAG]), 1990, various.

-Gondwana Eight: Assembly, Evolution, and Dispersal (A.A. Balkema, Rotterdam), 1993, R.H. Findlay, R. Unrug, M.R. Banks, and J.J. Veevers (eds).

-Initial Reports of the Deep Sea Drilling Project (US Government Printing Office, Washington DC), various.

-Marine Geomorphology, (X-libris Corp., Philadelphia USA), 310 p, 2001, N.C. Smoot, D.R. Choi, and M.I. Bhat.

-New Concepts in Global Tectonics, (Texas Tech University Press, Lubbock), 1992, S. Chatterjee and N. Hotton, III (eds).

-New Concepts in Global Tectonics Newsletter, Dong Choi and J.MacGregor Dickins (eds).

-The Origin and Evolution of Seamounts, Journal of Geophysical Research, Vol. 89, No. B13, 1984, A.B. Watts (ed).

-Phanerozoic faunal and floral realms of the earth: the intercalary relations of the Malvinokaffric and Gondwana faunal realms with the Tethyan faunal realm, Geological Society of America Memoir 189, 78 p., 1996, A.A. Meyerhoff, A.J. Boucot, D. Meyerhoff-Hull, and J.M. Dickins.

-Proceedings of the International Symposium on New Concepts in Global Tectonics, 1998, L. Maslov (ed).

-Scientific Events Alert Network Bulletin (Smithsonian, Washington DC), various.

-Special Issue: Geophysical and Geochemical Studies of the Izu-Bonin-Mariana Arc System, The Island Arc, Vol. 7, No. 3, 1998, R.J. Stern and M. Arima (eds).

-Special Volume on New Concepts in Global Tectonics, Himalayan Geology, Vol. 22, No. 1, 2001, J.M. Dickins, A.K. Dubey, D.R. Choi, and Y. Fujita (eds).

-Special Volume 5 on Earth Dynamics Beyond the Plate Paradigm, Bollettino della Societe Geolologica Italiana, 2005, F.C. Forese (ed).

-Surge Tectonics: A New Hypothesis of Global Tectonics, Kluwer Academic Publishers, Dordrecht), 1996, D. Meyerhoff-Hull (ed).

-Tectonic Globaloney, (X-libris Corp., Philadelphia USA), 165 p. 2003, N.C. Smoot.

APPENDICES

A. Geologic Time Scale
B. Earthquake Epicenters and Hypocenters
C. Seismic Stratigraphy Reinterpreted”

1999 GEOLOGIC TIME SCALE

CENOZOIC

Period		Epoch		Age	Picks (Ma)
Quaternary		Holocene			0.01
		Pleistocene		Calabrian	1.8
Tertiary	Neogene	Pliocene	L	Piacenzian	3.6
			E	Zanclean	5.3
		Miocene	L	Messinian	7.1
				Tortonian	11.2
			M	Serravallian	14.8
				Langhian	16.4
			E	Burdigalian	20.5
				Aquitanian	23.8
	Paleogene	Oligocene	L	Chattian	28.5
			E	Rupelian	33.7
		Eocene	L	Priabonian	37.0
			M	Bartonian	41.3
				Lutetian	49.0
			E	Ypresian	54.8
		Paleocene	L	Thanetian	57.9
				Selandian	61.0
			E	Danian	65.0

MESOZOIC

Period	Epoch		Age	Picks (Ma)	Uncert. (m.y.)
Cretaceous	Late			65	.2
			Maastrichtian	71.3	1
			Campanian	83.5	1
			Santonian	85.8	1
			Coniacian	89.0	1
			Turonian	93.5	4
			Cenomanian	99.0	1
	Early		Albian	112	2
			Aptian	121	3
		Neocomian	Barremian	127	3
			Hauterivian	132	4
			Valanginian	137	4
			Berriasian	144	5
Jurassic	Late		Tithonian	151	6
			Kimmeridgian	154	7
			Oxfordian	159	7
	Middle		Callovian	164	8
			Bathonian	169	8
			Bajocian	176	8
			Aalenian	180	8
	Early		Toarcian	190	8
			Pliensbachian	195	8
			Sinemurian	202	8
			Hettangian	206	8
Triassic	Late		Rhaetian	210	8
			Norian	221	9
			Carnian	227	9
	Middle		Ladinian	234	9
			Anisian	242	9
	Early		Olenekian	245	9
			Induan	248	10

Rapid polarity changes

PALEOZOIC

Period		Epoch	Age	Picks (Ma)
Permian		L		248
			Tatarian	252
			Ufimian-Kazanian	256
			Kungurian	260
		E	Artinskian	269
			Sakmarian	282
			Asselian	290
Carboniferous	Pennsylvanian	L	Gzelian (S.)	296
			Kasimovian	303
			Moscovian (W.)	311
			Bashkirian (N.)	323
	Mississippian		Serpukhovian	327
		E	Visean	342
			Tournaisian	354
Devonian		L	Famennian	364
			Frasnian	370
		M	Givetian	380
			Eifelian	391
		E	Emsian	400
			Praghian	412
			Lockhovian	417
Silurian		L	Pridolian	419
			Ludlovian	423
			Wenlockian	428
		E	Llandoverian	443
Ordovician		L	Ashgillian	449
			Caradocian	458
		M	Llandeilian	464
			Llanvirnian	470
		E	Arenigian	485
			Tremadocian	490
Cambrian*		D	Sunwaptan*	495
			Steptoean*	500
		C	Marjuman*	506
			Delamaran*	512
		B	Dyeran*	516
			Montezuman*	520
		A		543

PRECAMBRIAN

Eon	Era	Bdy. Ages (Ma)
		543
Proterozoic	Late	900
	Middle	1600
	Early	2500
Archean	Late	3000
	Middle	3400
	Early	3800?

GEOLOGICAL SOCIETY OF AMERICA

 Compilers: A. R. Palmer, John Geissman

*International ages have not been established. These are regional (Laurentian) only. Boundary Picks were based on dating techniques and fossil records as of 1999. Paleomagnetic attributions have errors, Please ignore the paleomagnetic scale.

Sources for nomenclature and ages: Primarily from Gradstein, F., and Ogg, J., 1996, *Episodes*, v. 19, nos. 1 & 2; Gradstein, F., et al., 1995, SEPM Special Pub. 54, p. 95–128; Berggren, W. A., et al., 1995, SEPM Special Pub. 54, p. 129–212; Cambrian and basal Ordovician ages adapted from Landing, E., 1998, *Canadian Journal of Earth Sciences*, v. 35, p. 329–338; and Davidek, K., et al., 1998, *Geological Magazine*, v. 135, p. 305–309. Cambrian age names from Palmer, A. R., 1998, *Canadian Journal of Earth Sciences*, v. 35, p. 323–328.

APPENDIX B-1

APPENDIX B-2

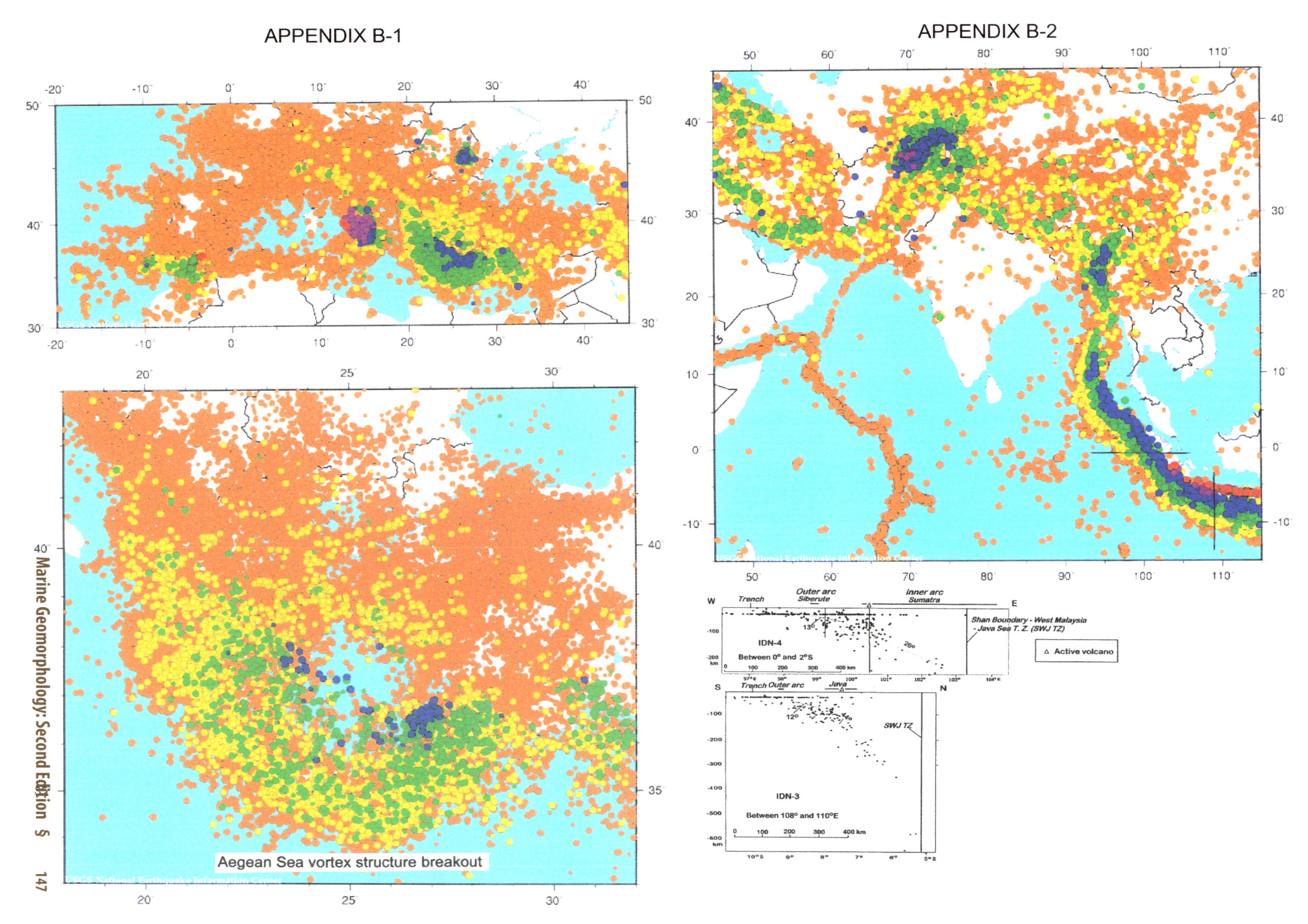

APPENDIX B-3

DEPTH
0
-33
-70
-150
-300
-500
-800
USGS National Earthquake Information Center
W
E
Shikoku Basin
Shichito-Iwojima Ridge
Ogasawara Trench
28°
45°
75°
SLM TZ
35°
OM-5
OGASAWARA ISLANDS
(Between 30°N and 32°N)
Volcanic Ridge
Trench
38°
SLM TZ
OM-4
OGASAWARA ISLANDS
Between 23°E and 24°E
Ridge
Saipan
Trench
16°
28°
SLM TZ
84°
OM-2
MARIANA ISLANDS
(Between 14°N and 16°N)
West Mariana Ridge
Shikoku Basin
Mariana Trough
Mariana Ridge
Trench
36°
60°
SLM TZ
Vertical
OM-3
MARIANA ISLANDS
(Between 18°N and 20°N)

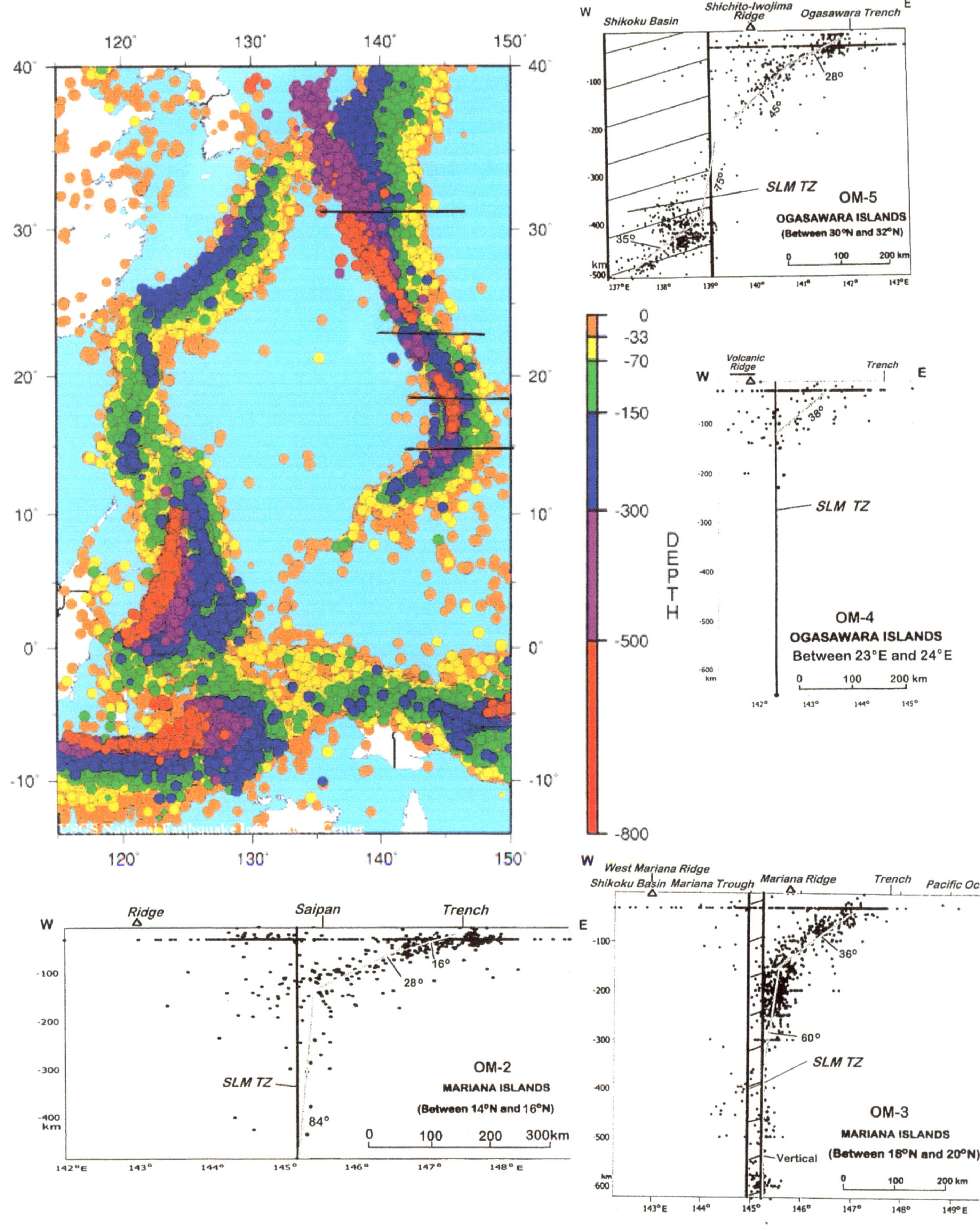

APPENDIX B-4

N
High heat flow
Duff Island
Malakula
New Hebrides Trench
Loyalty Islands
Lifou
Mare
S
-200
-400
km
CPM
Udintsev F. Z.
(Between 167°E and 168°E)
0
100
200
300
400
500 km
10° S
12°
14°
16°
18°
20°
22° S
15°
25
35
Loyalty Island
New Hebrides Trench
Hunter Ridge
High heat flow
North Fiji Basin
Hunter Ridge
South Fiji Basin
Lau Ridge
Tonga Ridge
Tonga Trench
W
E
ASEISMIC ZONE
-200
Udintsev F. Z.
-400
Fiji-Hunter Ridge Fault
Kermadec-Tonga Fault
(Between 21°S and 22°S)
Heezen F. Z.
-600
km
0
100
200 km
N-F. T. Z.
p7int.jpg
168° E
170°
172°
174°
176°
178°
180°
178°
176°
174°W
Colville Ridge
Kermadec Ridge
Kermadec Trench
Havre Trough
W
E
LOW SEISMICITY ZONE
-100
-200
-300
N-F. T. Z.
Udintsev F. Z.
-400
Kermadec-Tonga Fault
(Between 34°S and 35°S)
km
-500
0
100
200 km
177° E
178°
179°
180°
179°
178°
177°W
165°
175°
175°

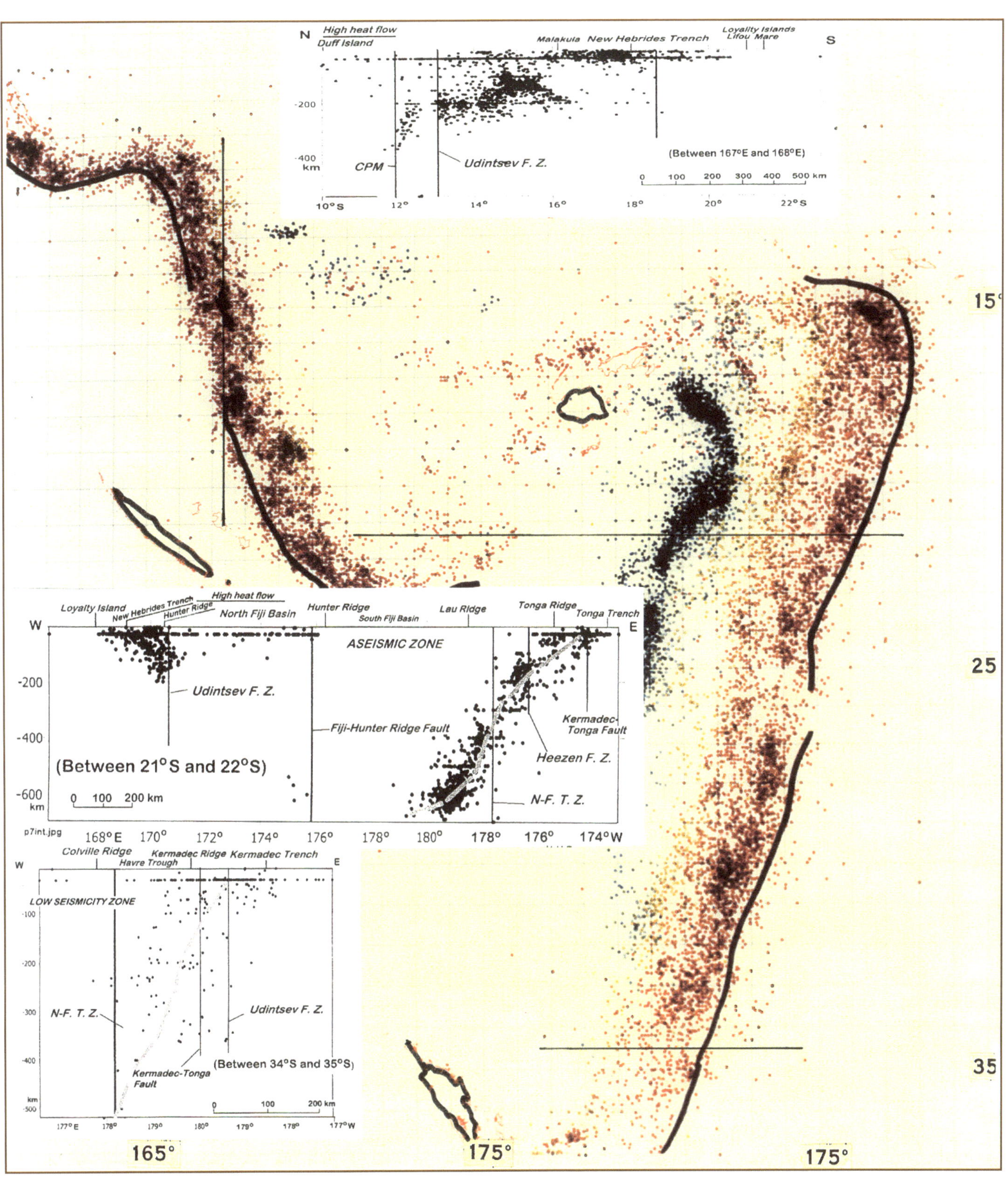

APPENDIX B-5

140°
150°
160°
170°
180°
60°
50°
40°
USGS National Earthquake Information Center
Seamount
Trench
Shelf
Coast
Alaska
Precambrian platform
Mesozoic - Cenozoic mobile belt
S
N
0
-100 km
150
240
Denali Fault
F-Q Fault
AL-1
Between 148°W and 150°W
0
500 km
56°N
57°
58°
59°
60°
61°
62°
63°
64°
65°N
60°
50°
USGS National Earthquake Information Center
180°
-170°
-160°
-150°
-140°
-130°
-120°

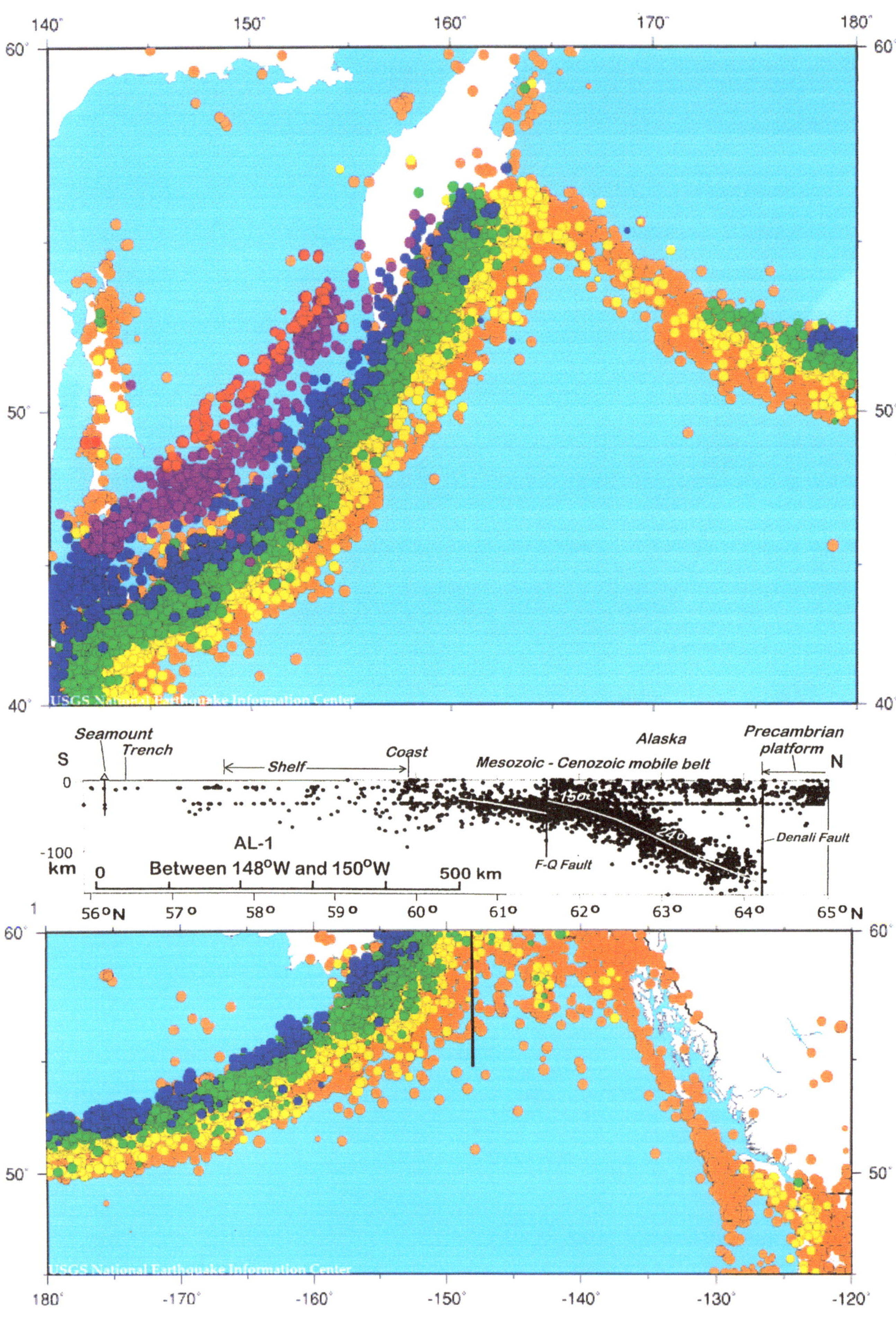

APPENDIX B-6

-140°
-130°
-120°
-110°
-100°
-90°
-80°
50°
40°
30°
20°
10°
USGS National Earthquake Information Center

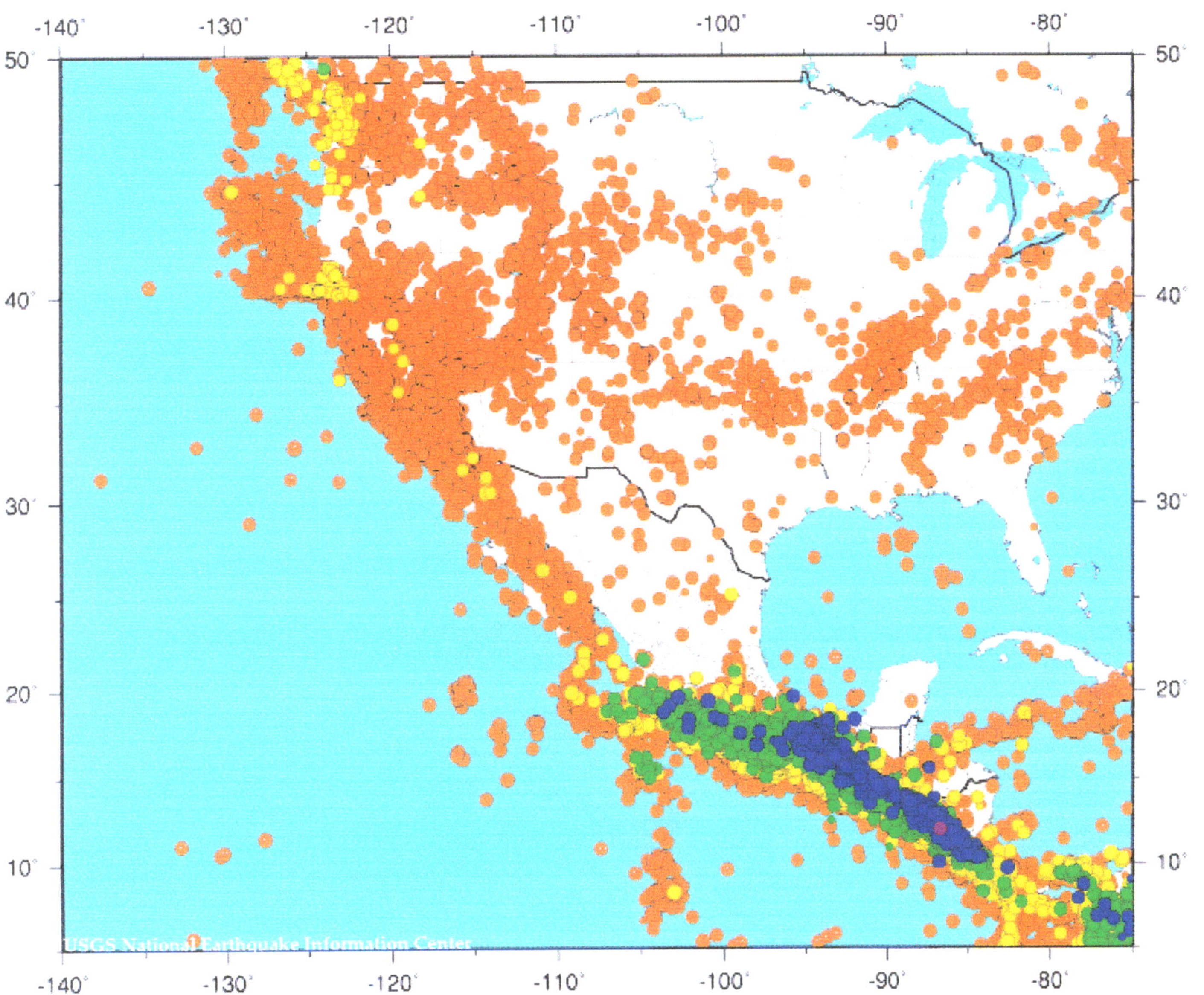

APPENDIX B-7

SAM-1
Between 5°S and 6°S
Mobile Belt
Brazilian Shield
Western Brazilian Shield Tectonic Zone (WBS TZ)
SAM-2
Between 8°S and 10°S
M.-C. Basin
Mesoz - Cenoz. Basin
SAM-3
Between 13°S and 14°S
SAM-4
Between 16°S and 18°S
Mesoz - Cenoz.
SAM-5
Between 21°S and 22°S
SAM-6
Between 26°S and 28°S
WBS TZ
Trench
Coast
Altiplano (3 km or higher)
Volcano
Pz1 Lower Paleozoic
PC Precambrian
USGS National Earthquake Information Center

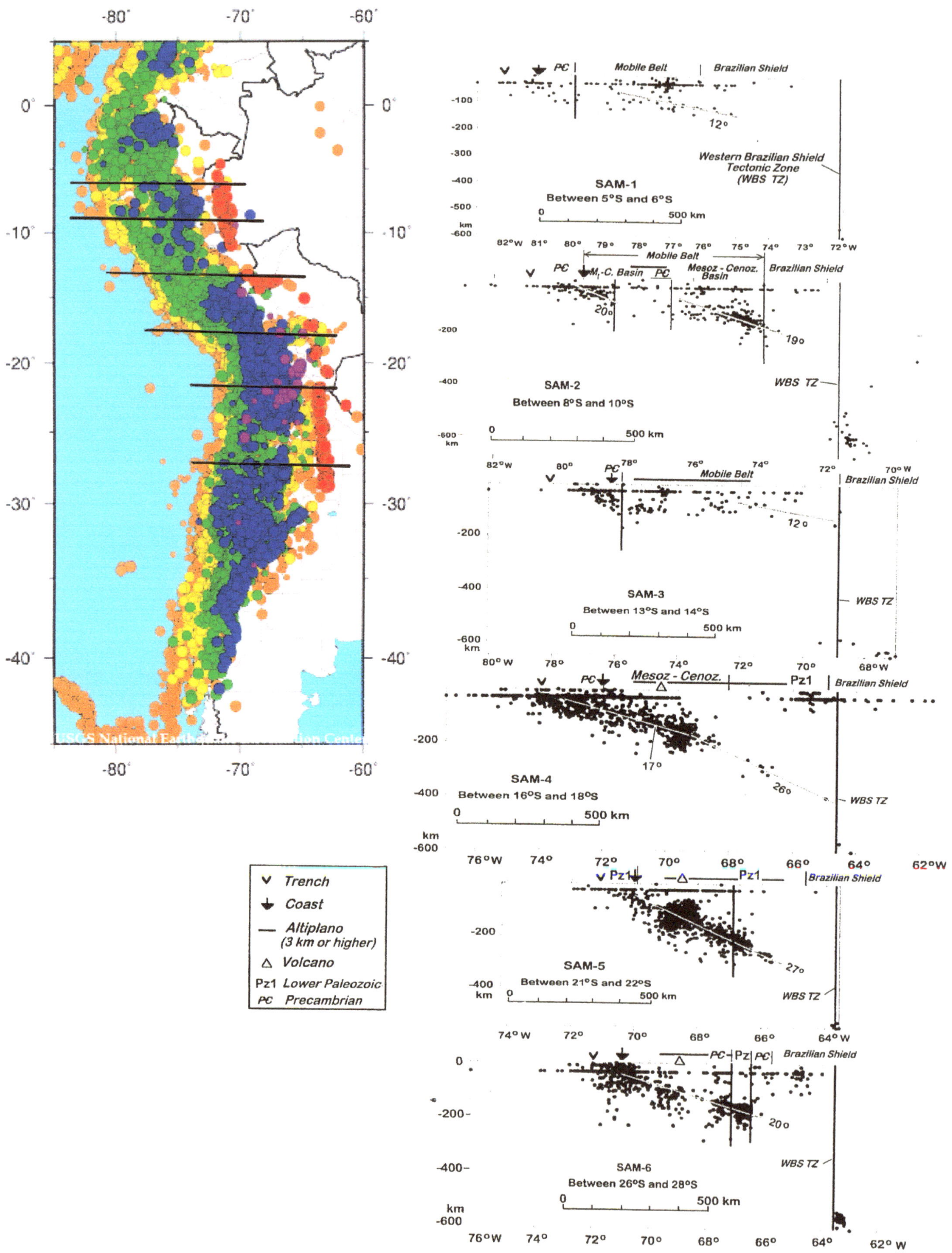

APPENDIX C-1

Aleutian Trench (~170°W)
N
S
Schist
MANTLE?
V : H = 1 : 1
Kuril Trench (~50°N)
W
E
V : H = 1 : 1
Japan Trench (~40°N)
LINE 78-4
Granodiorite clast
MANTLE
V : H = 1:1
0 10 20 30 km
Line GUA-18 off Guatemala (~14°N)
S
N
V : H = 3.6 : 1
0 5 10 km
Neogene-Quaternary
Basalt
Pre-Devonian (Proterozoic-Lower Paleozoic)
Paleozoic?
Jurassic? - Cretaceous - Paleogene
Archean?
Proterozoic-Lower Paleozoic
Line UTIG CR-20 off Costa Rica
SW
NE
Thrust zone
Plate interface thrust by von Huene (2003)
Pliocene - Quaternary
Pliocene - Pleistocene
Downlapping
Late Miocene
Early - Middle Miocene
Eocene?
Middle Miocene
Early Miocene
Gabbro
"Basement"
sec
1039
1040
1043
1042 (projected not to scale)
1041

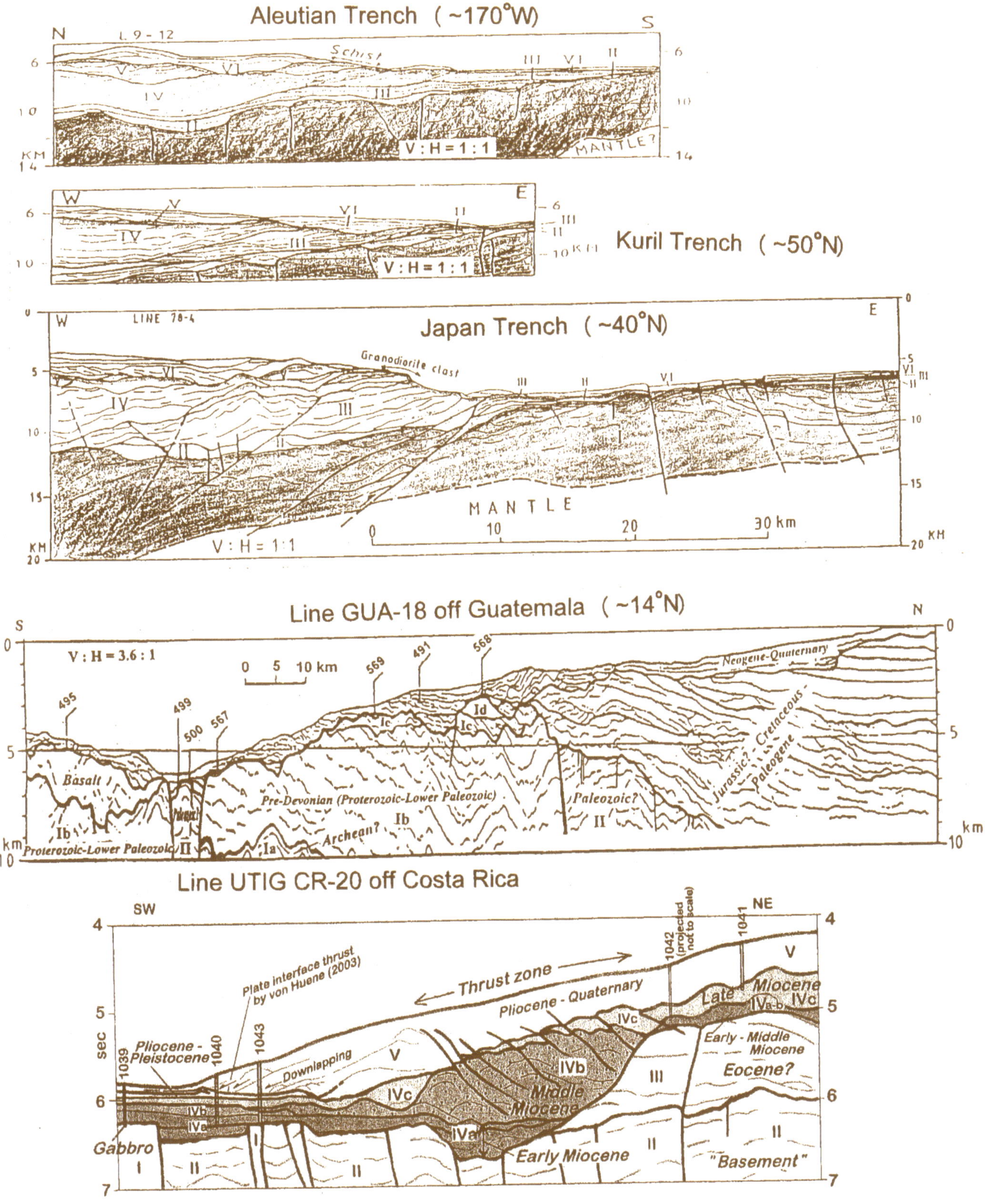

APPENDIX C-2

MIDDLE AMERICA TRENCH
W
E
495
BASALT
499
500
567
569
497
568
MESOZOIC TO CENOZOIC
I & II
III
PRECAMBRIAN
I & II
III
PALEOZOIC?
10 km
PRECAMBRIAN
I & II
ARCHEAN?
0 10 20 km
V : H = 1 : 1

W
Peru Trench (~14°S)
E
683
SEA LEVEL
PLIOCENE TO QUATERNARY
685
PLIOCENE-QUATERNARY
MIOCENE
Trench
BASALT
III
THRUST
MIOCENE
PALEOGENE
V
PALEOGENE
PALEOGENE?
IV
IV
PALEOZOIC TO MESOZOIC?
IV
I
PRECAMBRIAN
III PALEOZOIC
10 km
III
V : H = 1 : 1
0 10 km
LANDWARD SEDIMENT PROGRADATION

W
Chile Trench (45°S)
E
Trench
859
860
861
CENOZOIC
III
II
I
III
II
I
III
II
I
PALEOZOIC?
MESOZOIC?
V
PALEOZOIC?
IV
PRECAMBRIAN
III
V : H = 1 : 1
LANDWARD SEDIMENT PROGRADATION
I
0 10 KM
15 km
15km

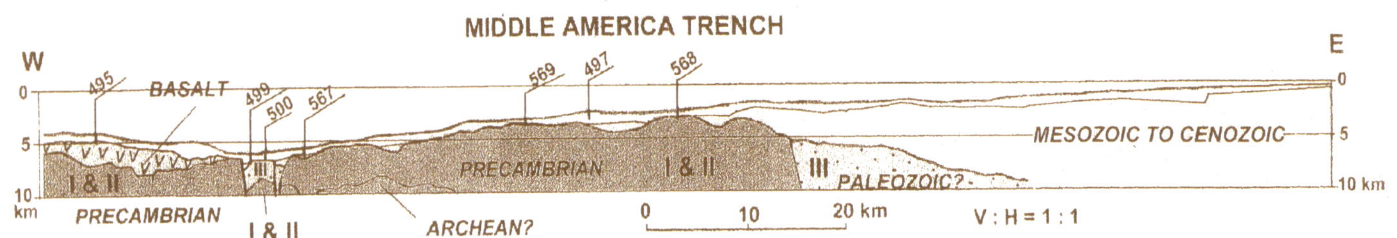

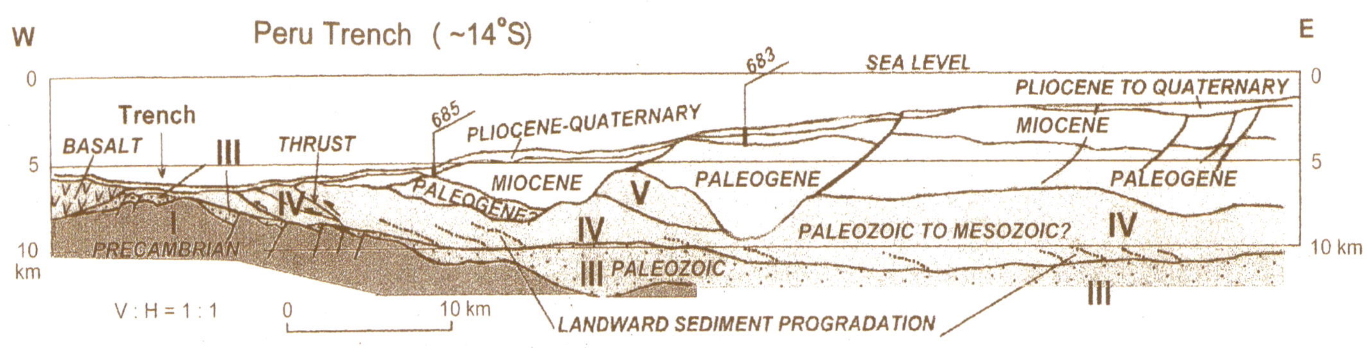

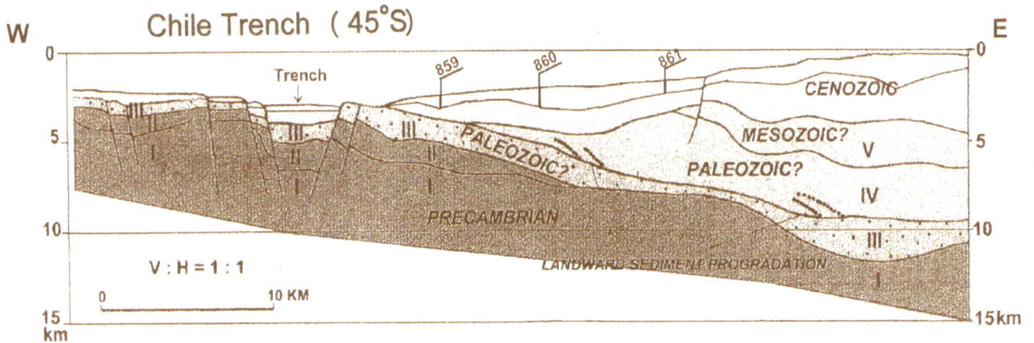

www.ingramcontent.com/pod-product-compliance
Ingram Content Group UK Ltd.
Pitfield, Milton Keynes, MK11 3LW, UK
UKHW060119300726
14090UKWH00002B/265
* 9 7 8 1 4 2 5 7 5 5 4 1 6 *